T0245479

CAMBRIDGE LIBRARY COLLECTION

Books of enduring scholarly value

Technology

The focus of this series is engineering, broadly construed. It covers technological innovation from a range of periods and cultures, but centres on the technological achievements of the industrial era in the West, particularly in the nineteenth century, as understood by their contemporaries. Infrastructure is one major focus, covering the building of railways and canals, bridges and tunnels, land drainage, the laying of submarine cables, and the construction of docks and lighthouses. Other key topics include developments in industrial and manufacturing fields such as mining technology, the production of iron and steel, the use of steam power, and chemical processes such as photography and textile dyes.

Memoirs of Richard Lovell Edgeworth, Esq

Richard Lovell Edgeworth (1744–1817) was a noted Irish educationalist, engineer and inventor. This two-volume autobiography, begun in 1808, was completed by his novelist daughter Maria, and published in 1820. Edgeworth's interest in education is evidenced by his reflections about how his childhood shaped his character and later life. Volume 1, written by Edgeworth himself and covering the period to 1781, reveals that his interest in science began early; he was shown an orrery (a moving model of the solar system) at the age of seven. As a young man, Edgeworth attended university in Dublin and Oxford, studied law, and eloped while still in his teens. He experimented with vehicle design, winning several awards, and was introduced by Erasmus Darwin to the circle of scientists, innovators and industrialists later known as the Lunar Society of Birmingham. In 1781 Sir Joseph Banks sponsored his election to the Royal Society.

Cambridge University Press has long been a pioneer in the reissuing of out-of-print titles from its own backlist, producing digital reprints of books that are still sought after by scholars and students but could not be reprinted economically using traditional technology. The Cambridge Library Collection extends this activity to a wider range of books which are still of importance to researchers and professionals, either for the source material they contain, or as landmarks in the history of their academic discipline.

Drawing from the world-renowned collections in the Cambridge University Library, and guided by the advice of experts in each subject area, Cambridge University Press is using state-of-the-art scanning machines in its own Printing House to capture the content of each book selected for inclusion. The files are processed to give a consistently clear, crisp image, and the books finished to the high quality standard for which the Press is recognised around the world. The latest print-on-demand technology ensures that the books will remain available indefinitely, and that orders for single or multiple copies can quickly be supplied.

The Cambridge Library Collection will bring back to life books of enduring scholarly value (including out-of-copyright works originally issued by other publishers) across a wide range of disciplines in the humanities and social sciences and in science and technology.

Memoirs of Richard Lovell Edgeworth, Esq

Concluded by his Daughter

VOLUME 1

RICHARD LOVELL EDGEWORTH
MARIA EDGEWORTH

CAMBRIDGE
UNIVERSITY PRESS

CAMBRIDGE UNIVERSITY PRESS

Cambridge, New York, Melbourne, Madrid, Cape Town, Singapore,
São Paolo, Delhi, Dubai, Tokyo, Mexico City

Published in the United States of America by Cambridge University Press, New York

www.cambridge.org
Information on this title: www.cambridge.org/9781108026567

© in this compilation Cambridge University Press 2011

This edition first published 1820
This digitally printed version 2011

ISBN 978-1-108-02656-7 Paperback

Engraved by A. Cardon

R. L. EDGEWORTH.

1812.

Published March 30th, 1820 by R. Hunter, No. 72, St Pauls Church Yard, London.

MEMOIRS

OF

RICHARD LOVELL EDGEWORTH, ESQ

BEGUN BY HIMSELF

AND

CONCLUDED BY HIS DAUGHTER,

MARIA EDGEWORTH.

IN TWO VOLUMES.

VOL. I.

LONDON:

PRINTED FOR R. HUNTER,

Successor to Mr. Johnson,

NO. 72, ST. PAUL'S CHURCHYARD,

AND

BALDWIN, CRADOCK, AND JOY,

PATERNOSTER ROW.

1820.

J. M'Creery, Printer,
Took's Court, London.

INTRODUCTION.

I WRITE my life, not because it contains extraordinary adventures, any uncommon series of good or bad fortune, any instances of superior talents, or heroic virtues ; but, because from habits acquired in educating a large family, I can develope with some certainty the circumstances, which have formed my character, and influenced my conduct.

My beloved daughter, Maria, at my earnest request, has promised to revise, complete, and publish her father's life.

Were she to perceive any extenuation on the one hand, or exaggeration on the other, it would wound her feelings ; she would be obliged to alter, or omit, what she did not approve, and her affection for her friend and parent would be diminished :—can the public have a better surety than this, for the accuracy of these memoirs?

In relating the life of any man who has never lived in absolute obscurity, some of the history of others must be interspersed. To steer between the extremes of too much, or too little, of what is not strictly the history of the writer, is difficult. Some of my readers may think that I say too much of my friends; and some may blame me for having omitted many particulars, relative to distinguished characters, which might have amused the public. The candid critic will be aware of these difficulties, and will be most likely to absolve me.

A slight narrative of my own feelings and actions may instruct and entertain those, who are curious in discriminating human character; and it may farther be of use, to shew that education continues from the cradle to the grave, or, to speak more accurately, from infancy to dotage;—Public and private events bias our conduct at every period of life; and I believe, because I feel, that the memory, and every faculty of what are called the heart and understanding, may not only be preserved, but may be improved, by care and attention, even between sixty and seventy, as well as between forty and fifty years of age.

I now take leave of the world, which has been most indulgent to me, as a man, and as an author, and I take leave of the world with this declaration,—that, to speak the truth without harshness, is, in my opinion, the most certain way to succeed in every honorable pursuit.

Whoever chuses to follow what is not honourable, must adopt more suitable advice.

CHAPTER I.

My family came into Ireland in the reign
of Queen Elizabeth, about the year 1583;
they had been established, as I have been
told, at Edgeworth, now called Edgeware
in Middlesex. A younger son of the
family, Roger Edgeworth, was a monk;
his elder brother dying, the fortune of the
family became the property of a sister,
who married into the family of Brydges.
This Roger Edgeworth wrote a sermon*
against the reformers; but, being smitten
like his master, Henry the Eighth, with
the bright eyes of beauty; like him, after
having been a defender of the catholic faith,
he reformed, and married. His sons Edward and Francis came to Ireland, probably under the patronage of Essex and
Cecil, as those names have continued chris-
tian names in my family ever since.

* This sermon I saw in the library of Corpus Christi
College, Oxford.

Edward Edgeworth, who was beneficed
in Ireland by Queen Elizabeth, was bishop
of Down and Connor in the year 1593;
and, dying without issue, he left his fortune to his brother, Francis Edgeworth,
who was Clerk of the Hanaper, in 1619.
This gentleman, from whom I am lineally
descended, married an Irish lady, Jane
Tuite, a daughter of Sir Edmond Tuite,
Knight, of Sonna, in the county of West-
meath. She was very beautiful, and of an
ancient family. It happened, that being
once obliged to give place at church to
some lady whom she thought her inferior, she pressed her husband to take out
a Baronet's patent, which had been prepared for him. At this time these patents
were, as he expressed it, " more onerous
than honorable;" and he refused to comply
with his wife's request. The lady, waxing
wroth, declared she would never go again
to church—the gentleman ungallantly replied, that she might stay or go, wherever
she pleased. In consequence of this permission, which she took in the largest
sense, she attached herself to Queen Henrietta Maria, with whom she continued in

France, during the remainder of the Queen's
life.   Upon her husband's refusing the
Baronet's patent, she obtained it for her
brother, Sir Edmond Tuite.   She returned
to Ireland afterwards, at Queen Henrietta
Maria's death ; but she disregarded her
husband's family and her own, and laid
out a very large fortune, in founding a religious house in Dublin.

Her son, Captain John Edgeworth, married the daughter of Sir Hugh Cullum, of
Derbyshire.   He brought her to Ireland,
to his Castle of Cranallagh, in the county
of Longford.   He had by her one son.
Before the Irish rebellion broke out, in
1641, Captain Edgeworth, not aware of
the immediate danger, left his wife and
infant in the Castle of Cranallagh, while
he was summoned to a distance by some
military duty.   During his absence, the
rebels rose, attacked the castle, set fire to
it at night, and dragged the lady out, literally naked. She escaped from their hands,
and hid herself under a furze bush, till
they had dispersed.   By what means she
saved herself from the fury of the rebels, I
never heard; she made her way to Dublin,

thence to England, and to her father's house in Derbyshire. After the rebels had forced this lady out of the castle, and had set fire to it, they plundered it completely; but they were persuaded to extinguish the fire from reverence for the picture of Jane Edgeworth. Her portrait was painted on. the wainscoat, with a cross hanging from her neck, and a rosary in her hands. Being a catholic, and having founded a religious house, she was considered as a saint. The only son of Captain Edgeworth was then an infant, lying in his cradle. One of the rebels seized the child by the leg, and was in the act of swinging him round to dash his brains out against the corner of the castle wall, when an Irish servant, of the lowest order, stopped his hand, claiming the right of killing the little heretick himself, and swearing that a sudden death would be too good for him; that he would plunge him up to the throat in a bog-hole, and leave him for the crows to pick his eyes out. Snatching the child from his comrade, he ran off with it to a neighbouring bog, and thrust it into the mud; but, when the rebels had retired, this man, who had only pretended to join them, went back to the bog for the boy, preserved his life, and, contriving to hide him in a pannier under eggs and chickens, carried him actually through the midst of the rebel camp, safely to Dublin. This faithful servant's name was Bryan Ferral. His last descendant died within my memory, after having lived, and been supported always, under my father's protection. My father heard this story from Lady Edgeworth, his grandmother, and also from a man of 107 years of age, one Bryan Simpson, who was present when the attack was made on Cranallagh Castle, and by whom the facts were circumstantially detailed.

Mrs. Edgeworth, the daughter of Sir Hugh Cullum, lived but a few years after her return to her father's house in Derbyshire. Her husband, Captain John Edgeworth, had followed her to England. Some time after he was left a widower, he determined to return to reside in Ireland. On his way thither, he stopped a day at Chester, it being Christmas day. He went to the Cathedral, and there he was struck with the sight of a lady, who had a full blown rose in her bosom. This lady was Mrs. Bridgman, widow of Mr. Edward Bridgman, brother to Sir Orlando Bridgman, the Lord Keeper. As she was coming out of church, the rose fell at Captain Edgeworth's feet. The lady was handsome—so was the captain—he took up the rose and presented it with so much grace to Mrs. Bridgman, that, in consequence, they became acquainted, and were soon after married. They came over to Ireland. Captain Edgeworth had a son, as I have mentioned, by his former wife, and the widow Bridgman had a daughter, by her former husband. The daughter was heiress to her father's property. These young people fell in love with each other. The mother was averse to the match. To avoid the law against running away with an heiress, the lovers settled, that the young lady should take her lover to church behind her on horseback. Their marriage was effected. Their first son, Francis, was born before the joint ages of his father and mother amounted to thirty-one years.

After the death of Captain Edgeworth

and his wife, which happened before this young couple had arrived at years of discretion, John Edgeworth took possession of a considerable estate in Ireland, and of an estate in England, in Lancashire, which came to him in right of his wife; he had also ten thousand pounds in money, as her fortune. But they were extravagant, and quite ignorant of the management of money. Upon an excursion to England, they mortgaged their estate in Lancashire, and carried the money to London, in a stocking, which they kept on the top of their bed. To this stocking, both wife and husband had free access, and of course its contents soon began to be very low. The young man was handsome, and very fond of dress. At one time, when he had run out all his cash, he actually sold the ground plot of a house in Dublin, to purchase a high crowned hat and feathers, which was then the mode. He lived in high company in London, and at court. Upon some occasion, King Charles the Second insisted upon knighting him. His lady was presented at court, where she was so much taken notice of by the gallant Monarch, that she thought it proper to intimate to her husband, that she did not wish to go there a second time, nor did she ever after appear at court, though in the bloom of youth and beauty. She returned to Ireland. This was an instance of prudence, as well as of strength of mind, which could hardly have been expected from the improvident temper she had shewn at first setting out in life. In this lady's character there was an extraordinary mixture of strength and weakness. She was courageous beyond the habits of her sex in real danger, and yet afraid of imaginary beings. According to the superstition of the times, she believed in fairies. Opposite to her husband's Castle of Lissard, in Ireland, and within view of the windows, there is a mount, which was reputed to be the resort of fairies; and when Lady Edgeworth resided alone at Lissard, the common people of the neighbourhood, either for amusement, or with the intention of frightening her away, sent children by night to this mount, who by their strange noises, by singing, and the lights they shewed from time to time, terrified her exceedingly. But she did not quit the place. The mount was

called Fairy-Mount, since abbreviated into
Fir mount.*

Of the courage and presence of mind of
this Lady Edgeworth, who was so much
afraid of fairies, I will now give an instance. While she was living at Lissard,
she was, on some sudden alarm, obliged to
go at night to a garret at the top of the
house, for some gunpowder, which was
kept there in a barrel. She was followed
up stairs by an ignorant servant girl, who
carried a bit of candle without a candlestick,
between her fingers. When Lady Edgeworth had taken what gunpowder she
wanted, had locked the door, and was half
way down stairs again, she observed, that
the girl had not her candle, and asked
what she had done with it, the girl recollected and answered, that she had left it
" *stuck in the barrel of black salt.*" Lady
Edgeworth bid her stand still, and instantly

* *Firmount.* From which in after times the Abbé
Edgeworth, to whose branch of the family this part of the
estate descended, called himself M. de Firmont.

The Abbé was Lady Edgeworth's grandson. Her fifth
son, Essex Edgeworth, was the abbé's father.

returned by herself to the room where the gunpowder was; found the candle as the girl had described—put her hand carefully underneath it—carried it safely out, and when she got to the bottom of the stairs, dropped on her knees, and thanked God for their deliverance. This lady, with all her courage and virtue, had a violent temper, which brought on family quarrels between her and her husband, and her many sons: so that the very early marriage, which I have mentioned, turned out unhappily. She recurred continually to the large fortune which she had brought her husband, and complained of being treated with neglect. As he grew older, Sir John became more prudent as to money matters. He pushed his fortune at court, and having considerable talents both as a courtier and a soldier, he obtained various places of trust and profit, and at last divided a large landed property among eight sons, leaving also a handsome jointure for his widow. The jointure lay upon the part of his estate that went to his eldest son, Colonel Francis Edgeworth, who, con-

sequently, was not rich. Lady Edgeworth
lived till she was ninety.

Francis Edgeworth, her eldest son, was
my grandfather. He was a loyal man,
and a zealous protestant, so much so, that
he was called *Protestant Frank*. In his
youth, he raised a regiment for King William, which, when he had completed, he
gave up to his father, Sir John, who required it from him. A memorandum of
an intended grant from the crown, of three
thousand pounds, on account of the expense of raising this regiment, and as an
acknowledgment for the service, still remains (unpaid) among our family papers.
My grandfather became colonel of the regiment after his father's death. He was a
man of great wit and gaiety, fond of his
profession, quite a soldier, and totally re-
gardless of money. He married successively several wives. One of whom, an
English lady, was a widow Bradstone.
Again, as in a former instance, which I
have mentioned, the widow had a daughter,
and a beautiful daughter, by her first husband. This daughter, Miss Bradstone,

my father's half sister, married Thomas
Pakenham, father to the first, and grandfather to the present Lord Longford. Thus
we became connected with the Pakenham
family. Colonel Francis Edgeworth, besides being straitened in his circumstances,
by having for many years a large jointure
to pay to his mother, was involved in difficulties by his own taste for play. A taste,
which, from indulgence, became an irresistible passion. One night, after having
lost all the money he could command, he
staked his wife's diamond ear-rings, and
went into an adjoining room, where she
was sitting in company, to ask her to lend
them to him. She took them from her
ears, and gave them to him, saying, that
she knew for what purpose he wanted them,
and that he was welcome to them. They
were played for. My grandfather won
upon this last stake, and gained back all he
had lost that night. In the warmth of his
gratitude to his wife, he, at her desire,
took an oath, that he would never more
play at any game with cards or dice. Some
time afterwards, he was found in a hay

yard with a friend, drawing straws out
of the hayrick, and betting upon which
should be the longest!—As might be expected, he lived in alternate extravagance
and distress; sometimes with a coach and
four, and sometimes in very want of half a
crown. He left his affairs in such disorder at his death, that his son, my father,
who was then an orphan of but eight years
old, must have lost his whole property,
had not Mr. Pakenham, his guardian,
taken care of him and of it. Mr. Pakenham,
though related to my father only by the half
blood, was as kind to him, as it was possible for the most affectionate parent to be.
Perceiving that my father was of an uncommonly steady disposition, Mr. Pakenham advised him to go to the Temple, at
eighteen, instead of going to college. This
prudent counsel my father followed, and
by application to business, and by making
himself master of his own affairs, he recovered a considerable part of his estate,
which had been unjustly detained from
him by some of his own family. He told
me a singular detection of fraud in one of
the suits, in which he was engaged : a deed

was produced against him, which was witnessed by a very old man, who was brought
into court. His venerable aspect prepossessed the court strongly in favour of his
veracity: he said that he was an antient
servant of the Edgeworth family, and had
been accustomed to transcribe papers for
the gentleman who had executed the deed.
He began, by declaring, that he had foreseen from the particular circumstances of the
deed, which went to disinherit the heir of
the family, that the transaction might hereafter be brought into dispute; he had therefore, he said, privately put a sixpence under the seal of the deed, which would appear if the seal were broken. The seal was
broken in open court, and the sixpence
was found to be dated five years subsequent to the date of the deed!—The deed
being thus proved to be a forgery, my father
gained his suit.

In a few years, my father found himself
in easy circumstances; and in 1732 he
married Jane Lovell, daughter of Samuel
Lovell, a Welsh judge, who was son of Sir
Salathiel Lovell, that recorder of London,
who, at the trial of the seven bishops, in the

reign of James II, proved himself to be a
good man, though he was but an indifferent
lawyer. He lived to the age of ninety-four,
and had so much lost his memory, as to be
called the *obliviscor* of London. Of him
I have heard my father relate an anecdote,
which has been told of others :—a young
lawyer pleading before him was so rude as
to say, "Sir, you have forgotten the law;"—
He replied, " Young man, I have forgotten
more law, than you will ever remember."

My grandfather, the Welch judge, travelling over the sands near Beaumorris, as
he was going circuit, was overtaken by the
night, and by the tide : his coach was set
fast in a quick-sand ; the water soon rose
into the coach, and his register, and some
other attendants, crept out of the windows
and mounted on the roof, and on the coach-
box. The judge let the water rise to his
very lips, and with becoming gravity replied, to all the earnest entreaties of his attendants, " I will follow your counsel, if
you can quote any precedent for a judge's
mounting a coach box."

This anecdote was told to my father by
Lady Edgeworth, that widow of Sir John,

who lived till ninety, and who related to
him many curious anecdotes of the five
reigns during which she flourished. From
her traditions, and from letters and papers,
now in my possession, my father compiled
some manuscript memoirs, from which I
was tempted here to make further extracts,
illustrative of the manners of the times.
Thinking, however, that they would take
up more room than could properly be
spared in this narrative, I omit all which do
not immediately relate to my own family.

CHAPTER II.

———

[1744.

AFTER my father's marriage with Miss
Lovell, he retired from the profession of
the law, and became a country gentleman.
He had eight children; four of whom died
during their childhood. My elder brother,
Thomas, my elder sister Mary, and Margaret, a sister younger than myself, sur-
vived. I was born in Pierrepoint street,
Bath, in the year 1744. By some misman-
agement my poor mother, at my birth, lost
the use of her right side, a misfortune that
was the more severely felt by her, as she
had been a remarkably active person. From
a sprightly young woman, who danced and
rode uncommonly well, she in one hour became a cripple for life.

I was unfortunate in my nurses. Before
my mother, who was now helpless, detected
their mal-practices, I was nearly starved

to death by two women, whose names were
so strange that I remember them to this
hour, nurse *Self* and nurse *Evil*. As soon
as my mother discovered the cause of my
sufferings, a good stout nurse was provided
for me; but, possibly, subsequent disorders
of my stomach proceeded from this early
starving. When I was between two and
three years old, I was carried over, with
my father and mother, to Ireland, to their
house at Edgeworth-Town, in the county of
Longford. I remember distinctly several
small circumstances, which happened before I was four years old. This I notice,
because the possibility of remembering
from so early an age has been doubted.
When I was about five years old, I   as
taught my alphabet: I remember well the
appearance of my hornbook; and once I
was beaten for not knowing the word *in-
step*. I recollect as distinctly as if it happened yesterday, that I had never before
heard or spelled that word. This un-
just chastisement put me back a little
in my learning; but as the injustice was
afterwards discovered, it saved me in
succeeding times from all violence from my

teachers. My mother then taught me to read herself. I lent my little soul to the business, and I read fluently before I was six years old. The first books that were put into my hands were the Old Testament, and Æsop's Fables. Æsop's Fables were scarcely intelligible to me: the frogs and their kings,—the fox and the bunch of grapes, confused my understanding; and the satyr and the traveller appeared to me absolute nonsense. I understood the lion and the mouse, and was charmed with the generous conduct of the one, and with the gratitude of the other. When I began to read the Old Testament, the creation made a great impression upon my mind: I personified the Deity, as is usual with ignorance. A particular part of my father's garden was paradise: my imagination represented Adam as walking in this garden; and the whole history became a drama in my mind. I pitied Adam, was angry with Eve, and I most cordially hated the devil. What was meant by Adam's bruising the serpent's head, I could not comprehend, and I frequently asked for explanations. The history of Jo-

seph and his brethren I perfectly understood ; it seized fast hold of my imagination, and of my feelings. I admired and
loved Joseph with enthusiasm ; and I believe, that the impression, which this history
made upon my mind, continued for many
years to have an influence upon my conduct.

My only play-fellow in my early childhood was my youngest sister Margaret ; my
elder sister was four years older than I was.
The early attachment which was formed between my sister Margaret (now Mrs. John
Ruxton) and me, has been one of the most
constant sources of pleasure that I have
ever possessed. There was and is a great
resemblance in our tempers, and characters, and tastes. I know how highly I
praise myself in saying this, but it must be
true, or we could not through so many different scenes of life have preserved as perfect a friendship and affection for each other,
as ever existed between brother and sister.
We were constant play-fellows, and such
constant friends, that for much more than
half a century the most violent, indeed I
may say the only quarrel, that we ever

had, was upon the following important occasion.

The gardener gave us some playthings, made of rushes; the good-natured old man presented them to us with much complacency, and divided them with impartiality. A gridiron he gave to little Miss; to little Master, a grenadier's cap. Little Miss, however, was not pleased with the distribution; she insisted upon having the grenadier's cap, which, after some reluctance on Master's part, she obtained: but, after having strutted her little hour under this heroic accoutrement, she became covetous of the more useful implement, with which she had seen me amusing myself. I had fried the gold fish, that were caught in the lake of the pond of the Black Islands; and I had gone through a considerable part of the story of the prince half marble and half man, as I had lately read it in the Arabian Nights Entertainments. I was in the character of the Black Genius, exclaiming, " Fish! fish! do your duty"— when my sister insisted that she ought to be the cook. I told her there were men cooks, but not female grenadiers. We disputed; we grew angry; we proceeded to violence; a battle ensued, in which the grenadier's cap was beaten to pieces. Loud were the lamentations. My mother heard the disturbance; and, instead of what is commonly called *scolding* us, took pains to do justice between us, and brought us to reason and peace, by mildly pointing out the folly of our quarrel. It is often from disputes like these, that children learn the consequences of passion, and the danger of giving way to it; and it is by the impartial and judicious conduct of parents, on such seemingly trivial occasions, that they may begin to form the temper to habits of self-command. Of this sister of mine I may say, that she has an uncommonly good temper, and she is as little inclined to violence as any of the gentlest of her sex. My mother took various means early to give me honourable feelings and good principles; and by these she endeavoured to correct, and to teach me to govern, the violence of my natural temper. She was lame, and not able to subdue me by force: if I ran away from her when she was going to punish me, she could not follow and catch me; but she obtained such power over my mind, that she induced me to come to her to be punished whenever she required it. I resigned myself, and without a struggle submitted to the chastisement she thought proper to inflict. The consequence of this submission was my acquiring, if I may say so, the *esteem* as well as the affection of my mother. But she was not blind to my faults: she saw the danger of my passionate temper. It was a difficult task to correct it; though perfectly submissive to her, I was with others rebellious and outrageous in my anger. My mother heard continual complaints of me; yet she wisely forbore to lecture or punish me for every trifling misdemeanor; she seized proper occasions to make a strong impression upon my mind.

One day, my elder brother Tom, who, as I have said, was almost a man when I was a little child, came into the nursery where I was playing, and where the maids were ironing. Upon some slight provocation or contradiction from him, I flew into a violent passion; and, snatching up one of the box-irons, which the maid had just laid down,

I flung it across the table at my brother. He stooped instantly; and, thank God! it missed him. There was a red hot heater in it, of which I knew nothing till I saw it thrown out, and till I heard the scream from the maids. They seized me, and dragged me down stairs to my mother. Knowing that she was extremely fond of my brother, and that she was of a warm indignant temper, they expected that signal vengeance would burst upon me. They all spoke at once. When my mother heard what I had done, I saw she was struck with horror, but she said not one word in anger to me. She ordered every body out of the room except myself, and then drawing me near her, she spoke to me in a mild voice, but in a most serious manner. First, she explained to me the nature of the crime, which I had run the hazard of committing; she told me, she was sure that I had no intention seriously to hurt my brother, and did not know, that if the iron had hit my brother, it must have killed him. While I felt this first shock, and whilst the horror of murder was upon me, my mother seized the moment, to conjure me to try in future to command my passions. I remember her telling me, that I had an uncle by the mother's side who had such a violent temper, that in a fit of passion one of his eyes actually started out of its socket. "You," said my mother to me, " have naturally a violent temper: if you grow up to be a man without learning to govern it, it will be impossible for you then to command yourself; and there is no knowing what crime you may in a fit of passion commit, and how miserable you may in consequence of it become. You are but a very young child, yet I think you can understand me. Instead of speaking to you as I do at this moment, I might punish you severely; but I think it better to treat you like a reasonable creature. My wish is to teach you to command your temper; nobody can do that for you, so well as you can do it for yourself."

As nearly as I can recollect, these were my mother's words; I am certain this was the sense of what she then said to me. The impression made by the earnest solemnity with which she spoke never, during the whole course of my life, was effaced from my mind. From that moment I determined to govern my temper. The determinations and the good resolutions of a boy of between five and six years old are not much to be depended upon, and I do not mean to boast, that mine were thenceforward uniformly kept; but I am conscious, that my mother's warning frequently recurred to me, when I felt the passion of anger rising within me; and that both whilst I was a child, and after I became a man, these her words of early advice had most powerful and salutary influence in restraining my temper.

Of the further rudiments of my education I recollect only that I was taught arithmetic, and made expert in counting at the card table, when my father and mother used to play cribbage. The attention to teach me numbers was bestowed particularly, because my father, not being infected with that foolish pride, which renders parents averse to the idea of putting a son *into business* or *commerce*, destined me for a merchant. My elder brother, however, dying when I was but six years old, I

became an only son. The views of my
education changed, and my life was now
to be preserved by an encreased degree of
care and precaution. I cannot say that my
dear mother, though she was a woman of
superior abilities, made much use of her
judgment with respect to the management
of my health, or to the education of my
body. Her fondness for my elder brother
was soon transferred to me; and to prevent
whatever might endanger my health, continually occupied her thoughts. At this
time, the humoral pathology was the creed
of physicians, and of those well-meaning
ladies, who watch over the constitutions of
their children, and endeavour, by continual
preventives, to avert every approach of
disease. I was naturally strong and active;
but I was now obliged to take a course of
physic twice a year, every Spring and
Autumn, with nine days' potions of small
beer and rhubarb, to fortify my stomach,
and to kill imaginary worms. I was not
suffered to feel the slightest inclemency of
the weather; I was muffled up whenever I
was permitted to ride a mile or two on
horseback before the coachman; my feet

never brushed the dew, nor was my head ever exposed to the wind or sun. Fortunately, my mother's knowledge of the human mind far exceeded her skill in medicine.

She inspired me with a love of truth, a dislike of low company, and an admiration of whatever was generous. Fortunately for me, the few visitors who frequented our house seemed to join with her in a wish to instil generous sentiments. One lady in particular, who, as I observed, was treated by my mother with much respect, made a salutary impression upon me. She gave me Gay's Fables with prints, with which I was much delighted; and desired me to get by heart the fable of the Lion and the Cub. She explained to me the design of this fable, which was within the compass of my understanding. It gave me early the notion, that I ought to dislike low company, and to despise the applause of the vulgar. Some traits in the history of Cyrus, which was read to me, seized my imagination, and, next to Joseph in the Old Testament, Cyrus became the favorite of my childhood. My sister and I used to amuse ourselves with playing Cyrus at the court of his grandfather Astyages. At the great Persian feasts I was, like young Cyrus, to set an example of temperance, to eat nothing but water-cresses, to drink nothing but water, and to reprove the cup-bearer for making the king my grandfather drunk. To this day I remember the taste of those water-cresses; and for those who love to trace the characters of men in the sports of children I may mention, that my character for sobriety, if not for water-drinking, has continued through life.

At seven years old, I became very devout. I had read some of the New Testament, and some account of the sufferings of martyrs; these inflamed my imagination so much, that I remember weeping bitterly before I was eight years old, because I lived at a time when I had no opportunity of being a martyr. I however dared to think for myself—My father was about this time enclosing a garden; part of the wall in its progress afforded means for climbing to the top of it, which I soon effected. My father reprimanded me severely, and as no fruit was at that time ripe, he could not readily conceive what motive I could have, for taking so much trouble, and running so great a risk. I told him truly, that I had no motive but the pleasure of climbing. I added, that if the garden were full of ripe peaches, it would be a much greater temptation ; and that unless he should be certain that nobody *would* climb over the wall, he ought not to have peaches in the garden. After having talked to me for some time, he discovered that I had reasoned thus : if my father knows beforehand, that the temptation of peaches will necessarily induce me to climb over the garden wall; and that if I do, it is more than probable that I shall break my neck, I shall not be guilty of any crime, but my father will be the cause of my breaking my neck. This I applied to Adam, without at the time being able to perceive the great difference between things human and divine. My father, feeling that he was not prepared to give me a satisfactory answer to this difficulty, judiciously declined the contest, and desired me not to meddle with what was above my comprehension. I mention this, because all parents, who encourage their children to speak freely, often hear from them puzzling questions and observations; and I wish to point out, that on such occasions children should not be discouraged, but on the contrary, according to the advice of Rousseau, parents should fairly and truly confess their ignorance.

So strong were my religious feelings at this time of my life, that I strenuously believed, that if I had sufficient faith, I could remove mountains; and accordingly I prayed for the objects of my childish wishes with the utmost fervency, and with the strongest persuasion that my prayers would be heard. How long the fervor of this sort of devotion lasted I do not remember; but I suppose, that going to school insensibly allayed it.

Before I was ready for school other circumstances occurred, which had considerable influence on my future character. The first was my mother's reading to me some passages from Shakespeare's plays, marking the characters of Coriolanus and of Julius Cæsar, which she admired. The contempt which Coriolanus expresses for

the opinion and applause of the vulgar, for
" the voices of the greasy-headed multi-
" tude," suited well with that disdain for
low company, with which I had been first
inspired by the fable of the lion and the
cub. It is probable that I understood the
speeches of Coriolanus but imperfectly;
yet I know, that I sympathised with my
mother's admiration, my young spirit was
touched by his noble character, by his generosity, and above all, by his filial piety,
and his gratitude to his mother. My mother
took every occasion to strengthen the impression this character made on my mind.
She was herself of a generous and grateful
disposition, and any instance of gratitude
called forth her warm approbation. She
considered gratitude as the strongest mark
of a generous character: she thought, that
to give or to forgive is generous; but that
to remember and to return favors is more
difficult than to confer them. In conferring
benefits the will is free, the mind acts from
warm spontaneous exertion, a sense of superiority accompanies the action; but the
feeling of gratitude implies a sense of obligation, and perhaps of inferiority. There

are but few. who have been habituated to think that to pay a debt is more agreeable than to bestow a favor. But in the same degree that a full and warm return for benefits is uncommon, we are pleased with ourselves for fulfilling this duty; and in fact, whoever has been often grateful, must have experienced as much self-satisfaction as the most liberal man upon earth can feel in sharing his fortune, or in bestowing his exertions to promote the interests of others.

I have mentioned, that my mother, even while I was but a child of eight years old, was in the habit of treating me like a reasonable being. She began to point out to me the good or bad qualities of the persons whom we accidentally saw, or with whom we were connected. About this time, one of our relations, a remarkably handsome youth of eighteen or nineteen, came one day to dine with us; my father was from home, and I had an opportunity of seeing the manners of this young man. He was quite uninformed; my mother told me, that he had received no education, that he was a hard drinker, and that notwithstanding his handsome appearance, he

would be good for nothing. Her predic-
tion was soon verified. He married a
woman of inferior station, when he was
scarcely twenty. His wife's numerous
grown-up-family, father, brothers, and
cousins, were taken into his house. They
appeared wherever any public meeting
gave them an opportunity, in a handsome
coach with four beautiful grey horses ; the
men were dressed in laced clothes after the
fashion of those days, and his wife's relations lived luxuriously at his house for
two or three years. In that period of time,
they dissipated the fee-simple of twelve
hundred pounds a year, which, fifty years
ago, was equal at least to three thousand of
our present money. The quantity of claret
which these parasites swallowed was so
extraordinary, that when the accounts of
this foolish youth came before the chan-
cellor, his lordship disallowed a great part
of the wine-merchant's bill ; adding, that
had the gentleman's coach-horses drunk
claret, so much as had been charged could
not have been consumed. This wine-merchant however obtained a considerable
portion of the poor young man's estate,

in liquidation of the outstanding debt. The host had for some time partaken of the good cheer in his own house; but disease, loss of appetite, and want of relish for jovial companions, soon confined him to his own apartment, which happened to be over the dining parlour, where he heard the noisy merriment below. In this solitary situation, a basin of bread and milk was one day brought to him, in which he observed an unusual quantity of hard black crusts of bread. He objected to them, and upon inquiry was told, that they were the refuse crusts, that had been cut off a loaf, of which a pudding had been made for dinner. This instance of neglect and ingratitude stung him to the quick; he threw the basin from him and exclaimed, "*I deserve it.*" To be denied a crumb of bread in his own house, where his wife's whole family were at that instant rioting at his expense, "quite conquered him." He never held his head up afterwards, but in a few months died, leaving a large family totally unprovided with fortune, to the guidance of a mother, who kept them destitute of any sort of instruction. These circumstances, and especially the anecdote of the basin of bread and milk, made too deep an impression on my mind, tending to inspire me with too great scepticism as to the gratitude of mankind. This opinion has, however, been in some degree effaced by experience; and I am now persuaded, that more ingratitude arises from the injudicious conduct of benefactors, than from the want of proper feeling in those whom they have obliged.

When the affairs of my relation were at his death the subject of conversation, my mother observed to me, that the cause of all the misfortunes that befel him was an easiness of temper, that led him to yield to every creature who attempted to persuade him. She desired me to remember, that young men are led into a thousand inconveniences from a false shame, that prevents them from refusing to do as others do; that the character for good nature, which it is so common to admire and to imitate, is not always obtained by real acts of generosity, or by real feelings of good-will, but by mere undistinguishing easiness of temper—that people of this temper, well aware of their own weakness, join with apparent eagerness in every proposal which *shews spirit.* She observed, that where the cry of numbers hurries us on, we may shew good fellowship, and even sympathy; but we do not deserve to be applauded for *spirit.* Real spirit is shown in resisting importunity and examples, and in daring to do what we think right, independently of the opinion of others.

But to return to the history of my childhood.—When I was about seven years old, a circumstance happened, which had considerable effect in forming the principal taste of my life, though at first view it seemed to concern me but little. A gentleman and his wife, on their way to Dublin, were delayed by the sickness of the lady at a wretched inn at Edge-worth-Town. My mother not only sent what was proper, but invited the distressed travellers to her house, and took such effectual care of the sick lady, that in a few days she recovered, and pursued her journey. When my mother went some time afterwards to Dublin, she took me with her. Mr. Deane, the husband of the sick lady, came to see my mother; and, as he got out of his coach, I observed, that he had brought with him a nice mahogany table, and some uncommon pieces of machinery, which excited my curiosity not a little. These were the parts of an electrical machine, that Mr. Deane had made, and which he presented to my mother, in hopes that it might be of service in alleviating the effects of the palsy with which she was affected. The benevolent countenance, melodious voice, and grateful conduct of this gentleman, made a great impression on my young mind. I was permitted, after much entreaty, to be present whilst the experiment was going on. At this time electricity was but little known in Ireland, and its fame as a cure for palsy had been considerably magnified. It, as usual, excited some sensation in the paralytic limb on the first trials. One of the experiments on my mother failed of producing a shock, and Mr. Deane seemed at a loss to account for it. I had observed, that the wire, which was used to conduct the electric fluid, had, as it hung in a curve from the instrument to my mother's arm, touched the hinge of a table which was in the way, and I had the courage to mention this circumstance, which was the real cause of failure. Mr. Deane was so well pleased with my observation, that he took me up in his arms, kissed me, and invited me to come the next morning to see his study and his workshop. I was sent there at the hour appointed, and the good-natured philosopher condescended to answer a number of questions, which my eager curiosity suggested. The apartment and its contents are now present to my memory, though it is near sixty years since I was there. Mr. Deane was then making an orrery, which he afterwards bequeathed to the University of Dublin. This orrery instantly caught my attention, but as its uses could not be explained to me, he very wisely turned my attention to another object, and shewed me the engine for cutting teeth in clock-wheels. He was then finishing some large wheels for his orrery, and he explained the parts and the uses of the engine so clearly, that I soon understood them. He then showed me a large globe, with which I was much pleased. I had been used to examine the map of the world upon a great screen at home; but though I had seen a map, I had not seen a globe, and I recollect, that when I found Italy, Sicily, and Ireland, upon the globe, I was delighted with the new idea which I received of the relative situation of places on the earth. Mr. Deane shewed me the use of several tools, which are employed by the makers of mathematical instruments; he shewed me a syphon, and the parts of a clock; he melted some metal for me in a crucible; he explained to me the bellows of an organ pipe, and many other mechanical contrivances. He bestowed praise upon my attention, and upon what he was pleased to call my intelligence; so that from the pleasure I received, and the impression made upon my mind that morning, I became irrecoverably a mechanic. These are circumstances in themselves so trifling, that I should not think of relating them, were it not to shew in one instance, at least, the truth of what I have elsewhere asserted, that what is usually called in children a genius for any particular art or science is nothing more than the effect of some circumstance, that makes an early impression, either from a strong association of pleasure or pain: such circumstances are most commonly accidental; but sometimes they are purposely thrown in the way, to produce a particular propensity in youth.

CHAPTER III.

———

[1752.

ONE of the great eras in a boy's life now
approached;—an introduction to the rudiments of ancient learning.—A clergyman
of the neighbouring village was engaged
to teach me the latin grammar. This gentleman was the Reverend Patrick Hughes,
who had the honor of leading Goldsmith
and some other conspicuous literary characters over the threshold of learning. He
put into my hands, without any previous in-
auguration, that universal book of knowledge, Lilly's Grammar. Never can I forget
the amazement, or rather the stupor, which
overwhelmed my mind, when I read, and
attempted to learn by heart, the first de-
finition in this learned manual. By pure
dint of reiteration, I got my lessons with
tolerable success ; but I am satisfied, that

my pedagogue augured ill of the young
squire. I, however, like other boys, drudged
on through thick and thin. I got through
Tristis (sad) without any misfortune ; but
the first and only punishment I ever suffered in the cause of classical literature
was occasioned by *Felix* (happy), and I
well recollect the cause of my failure and
disgrace. While I learned my lesson by
rote, I had the trick of standing upon
one leg, buckling and unbuckling my shoe.
When I came to repeat my lesson, my
master insisted upon my standing still and
on both legs. My memory was directly
at a fault. My associations were broken,
and I could not go on, through *hic hæc et
hoc felix*, without taking up my forbidden
foot. A *good* whipping, as it is called,
cured me of the trick, and of what *appeared
to be* obstinacy or stupidity. After a few
months' preparatory discipline, I was taken
to school at Warwick, and placed under
the care of Dr. Lydiat, on the 26th of August, 1752, a day memorable to me. A
new world now opened to my view. I was
about eight years old; I had been bred up
with much tenderness, and had never be-

fore lived with any companions but my
sisters. The noise, and bustle, and roughness of my schoolfellows, at first con-
founded me; I had been sufficiently tainted
with Irish accent, and Irish idiom, to be
the object of much open ridicule, and much
secret contempt. I beat one boy, who was
taller than myself, for mocking me; and
in a short time I acquired the English
provincial accent of my companions so
effectually, as to give no fair pretence for
tormenting me on this subject; but I still
retained the name of *Little Irish,* which
would have continued to be my nick-name,
but for a slight circumstance, of which I
made happy advantage. The races at
Warwick inflamed my companions with
the love for racing. A famous match, I
believe in the year 1752, between *Little-
witch* and *Little-driver,* occupied the attention of all the grown and growing children
of the neighbourhood. One of the cap-
tains of the school, my worthy friend Frank
Mundy, of Markeaton, in Derbyshire,
matched me to run against a boy, my superior in age, and famous for agility. Fortunately, this boy was for the occasion

named *Little-witch*; I declared, that I would not run, unless the name of " *Little-driver*" was given to me. I won the race, and the name of *Little-driver* necessarily superseded that of *Little Irish*. Nicknames are disagreeable appendages, and care should be taken, to prevent or shake them off. I have known a young man of eminence kept for a long time below his level amongst his companions, from his having succeeded to a nickname that his elder brother had acquired, from some circumstance in his conduct, which was in no manner whatever connected with the younger brother's character, or with the real estimation in which he was held by his schoolfellows. At Warwick, I learned not only the first rudiments of grammar, but also the rudiments of that knowledge, which leads us to observe the difference of tempers and characters in our fellow-creatures. The marking how widely they differ, and by what minute varieties they are distinguished, continues to the end of life, an inexhaustible subject of discrimination!

I had been accustomed to the affection of all my family at home, and was totally unacquainted with that love of power and of tyranny, which seems almost innate in certain minds. A full grown boy, just ready for college, made it his favourite amusement, to harass the minds, and torment the bodies, of his younger school-fellows. A little boy, with remarkably long flaxen hair, and myself, were the chosen objects of his cruelty; he used to knot our hair together, and drag us up and down the school-room stairs, for his diversion. One evening, when Dr. Lydiat, and all the boys, excepting my tormentor and myself had gone to church, he caught me, and confining me with iron grasp between his knees, he pulled a small black box from his pocket, which, with a terrific voice and countenance, he informed me was filled with dead men's fat; with the fat of a man who had lately been hanged; this he invited me to eat, and upon my refusing to do so with manifest signs of horror and disgust, he crammed my mouth with it till I was nearly suffocated.—The box contained, it is true, nothing but spermaceti, but to me it was dreadful as poison. A few days afterwards, when the tart woman came, he again seized me, and again attempted to cram my mouth with the contents of his accursed box, instead of permitting me to regale myself with damson tart. Roused to desperate resistance, I struck him in the face with my utmost strength; he, of course, knocked me down so decidedly, as to make it doubtful whether I should ever get up again. Another schoolfellow of mine was present, Christopher Wren, grandson of the great Sir Christopher. Though far inferior in strength and size to my tormentor, Wren could not restrain his indignation from venting itself in terms, that immediately produced a blow. A battle ensued, which would, as the spectators said, have terminated in favor of my champion, if it had continued; but I had run into the room where Dr. Lydiat was, an action of no common daring, and informed him of the combat. Wren met with the applause which was due to his humanity and courage, not only from his master, but from his schoolfellows. From me he won my warmest affection, which never ceased while he lived, and

E 2

which, in this sixty-fifth year of my life, continues so strongly impressed on my memory, that I feel a prepossession in favor of every person of his name—a prepossession, which has never disappointed me, whenever I have become acquainted with any of the descendants of our immortal architect. That one of the most amiable virtues of human nature was possessed by his son, is proved by the publication of Parentalia, in which the history of the author's father and of his works is given with internal evidence of truth, with the highest reverence for his ancestor, and yet without a single circumstance of exaggeration. Every reader, who has at any period of his life received protection from injustice and oppression, will sympathise with my feelings of gratitude.

My father and his family were at Bath when my first Christmas holidays approached. Travelling in England was at that time very different from what it. is at present; to send to Warwick for me, and to convey me back again, was inconvenient. I, therefore, must have remained at school,

had not Mrs. Dewes*, sister to Mrs. Delany, taken compassion upon me. Mrs.
Dewes lived at Welsbourne, within four
miles of Warwick, where her four sons
were at school with me. The youngest
was my class fellow, and upon a visit,
which Mrs. Dewes paid to Dr. Lydiat,
she saw me, and wrote to my mother at
Bath, to obtain her consent to my passing
the holidays at Welsbourne. Fortunately,
some former connexion, which had subsisted between my mother and Mrs. Delany, smoothed all difficulties; I was permitted to accept the invitation, and I was
received by Mrs. Dewes with the utmost
kindness and cordiality. Old English hospitality never appeared to me to be exercised with more propriety than at Welsbourne. The tenants of Mr. Dewes were
invited to a Christmas dinner of excellent
cheer, and their wives and daughters passed the evening in mirth and *unreproved
pleasure.* The fiddle and a good supper
sent all the young people happy to their

---

* This lady's name is mentioned with eulogium in Mrs
Barbauld's preface to Richardson's works.

homes; and Mrs. Dewes's cheerful and instructive conversation spread universal satisfaction among the elder part of the company. This scene is still present to my recollection. Three Miss Shendalls, tall,
handsome girls from Stratford, were in-
mates for some days at Welsbourne. They
were obviously superior to the generality
of the visitors; their conversation with Mrs.
Dewes was generally upon what they were
reading; and as they sat round a table
every evening intent upon their books, my
young friend and I, from imitation, employed ourselves in perusing certain little
volumes, which were then printed by New-
bery, for children, or we deciphered ana-
grams, which Mrs. Dewes and her friends
gave us for our amusement. To this occupation we were much encouraged by the
attention, which the elder part of the company gave to the subject, when a difficult
combination of letters fell in their way.
Upon such slight circumstances as these
the first tastes for different pursuits in life
are formed, and to a number of concurring
circumstances of this sort I owed an early

taste for reading, which continued with but little interruption during the whole course of my life.

After my return to school, my progress in Latin was prevented by the hooping cough ; cut off from all occupation and amusement, my time passed heavily, and I should have been in a miserable situation, had I not been treated with the utmost kindness and tenderness by Dr. Lydiat and by his sister, who managed his house. My father soon came from Bath, and took me away with him. I parted from my young friends of the Dewes family, with regret; and I did my best to express my gratitude to Dr. Lydiat, and to his sister, Mrs. Gore.

Our journey lay in some places out of the high road, and across corn fields. Our vehicle was a two wheeled carriage, some-thing like a French *chaise de poste,* and as we travelled slowly, I had time for obser-vation. I recollect, however, only one thing that caught my attention : when we came on the high road to Cirencester, I saw a man carrying a machine five or six feet in diameter, of an oval form, and composed of slender ribs of steel. I begged my father to inquire what it was. We were told, that it was the skeleton of a lady's hoop. It was furnished with hinges, which permitted it to fold together in a small compass, so that more than two persons might sit on one seat of a coach, a feat not easily performed, when ladies were encompassed with whalebone hoops of six feet extent. My curiosity was excited by the first sight of this machine, probably more than another child's might have been, because previous agreeable associations had given me some taste for mechanics, which was still a little further increased by the pleasure I took in examining this glittering contrivance. Thus even the most trivial incidents in childhood act reciprocally as cause and effect in forming our tastes.

What passed from the time when I left Warwick, till my father's return to Ireland, I do not distinctly remember; it therefore passed most probably without any event, that had much effect upon my temper or understanding. I recollect only that my mother's health had declined after

her return from Bath. Every thing that
art could suggest was tried for her recovery.

About the year 1754, Lord Trimblestone, a Roman catholic nobleman, who
had resided many years abroad, had become famous for his skill in medicine, and
for his benevolent attention to persons of
all ranks who applied to him. Success
encreased his popularity. With his permission, my father and mother went to
Trimblestone to consult him. They took
up their lodgings in the neighbourhood of
Trim, near his seat, and were most cordially. and hospitably received. My mother's complaints his lordship did not hope
to cure; but he ordered such palliation as
the art of medicine affords. I remember
in particular, that the *uva ursi,* which has
been lately resorted to in nephritic complaints, was prescribed for her, and with
great difficulty obtained. My mother was
pleased with the manners, and appearance,
and mode of life at Trimblestone, and
when she returned home, though she had
received, and had indeed expected, but
little advantage from Lord Trimblestone's

prescriptions, she often related with pleasure, circumstances that passed while she
was at his house. At the time which I
speak of, most families in Ireland dined
at three or four o'clock, but Lord Trimblestone never dined till seven or eight in
the evening. This arrangement gave an
air of mystery to his Lordship's domestic
economy, which added perhaps to the influence of his real skill over the minds
of his patients. My mother was much
struck with the beauty and grace of Lady
Trimblestone. She had, I believe, passed the meridian of life: but the glare
of a profusion of light; the fine brilliant
pendants in her ears; the unchanged colour
of her beautiful hair, which fell exube-
rantly in ringlets upon her neck,—contrary
to the fashion of that day, which required
that a lady's hair should be powdered and
rolled over a cylinder of black silk stuffed
with wool, so as to draw the hair almost
out by the roots from the forehead;—and
above all, an air of ease and dignity which
she had acquired in France; made her, at
least to the eyes of my mother, a most
charming and interesting person. Lord

Trimblestone commanded attention from his high character in the world for medical knowledge and for philanthropy. They soon set their stranger guests at ease, and both of them amused their auditors with accounts of remarkable patients, diseases, and cures, which they had witnessed. One in particular I will relate.

A very delicate lady of fashion, who had, till her beauty began to decay, been flattered egregiously by one sex, and vehemently envied by the other, began to feel as years approached, that she was shrinking into nobody. Disappointment produces ennui, and ennui disease; a train of nervous symptoms succeeded each other with alarming rapidity, and after the advice and the consultations of all the physicians in Ireland, and the correspondence of the most eminent in England, this poor lady had recourse in the last resort to Lord Trimblestone. He declined interfering, he hesitated; but at last, after much intercession, he consented to hear the lady's complaints, and to endeavour to effect her cure: this concession was made upon a positive stipulation, that the pa-

tient should remain three weeks in his
house, without any attendants but those of
his own family, and that her friends should
give her up entirely to his management.—
The case was desperate, and any terms
must be submitted to, where there was a
prospect of relief. The lady went to Trimblestone, was received with the greatest attention and politeness. Instead of a grave
and forbidding physician, her host, she found,
was a man of most agreeable manners. Lady
Trimblestone did every thing in her power
to entertain her guest, and for two or three
days the demon of ennui was banished. At
length the lady's vapours returned; every
thing appeared changed. Melancholy
brought on a return of alarming nervous
complaints—convulsions of the limbs—perversion of the understanding—a horror of
society; in short, all the complaints that
are to be met with in an advertisement
enumerating the miseries of a nervous patient. In the midst of one of her most
violent fits, four mutes, dressed in white,
entered her apartment; slowly approaching, they took her without violence in
their arms, and without giving her time

to recollect herself, conveyed her into a
distant chamber hung with black, and
lighted with green tapers. From the
ceiling, which was of a considerable height,
a swing was suspended, in which she was
placed by the mutes, so as to be seated at
some distance from the ground. One of
the mutes set the swing in motion; and as
it approached one end of the room, she
was opposed by a grim menacing figure
armed with a huge rod of birch. When
she looked behind her, she saw a similar
figure at the other end of the room, armed
in the same manner. The terror, notwithstanding the strange circumstances which
surrounded her, was not of that sort which
threatens life; but every instant there was
an immediate hazard of bodily pain. After
some time, the mutes appeared again,
with great composure took the lady out of
the swing, and conducted her to her apartment. When she had reposed some time,
a servant came to inform her, that tea was
ready. Fear of what might be the consequences of a refusal prevented her from
declining to appear. No notice was taken
of what had happened, and the evening

and the next day passed without any attack of her disorder. On the third day the vapours returned—the mutes reappeared—the menacing flagellants again affrighted her, and again she enjoyed a remission of her complaints. By degrees the fits of her disorder became less frequent, the ministration of her tormentors less necessary, and in time, the habits of hypochondriacism were so often interrupted, and such a new· series of ideas was introduced into her mind, that she recovered perfect health, and preserved to the end of her life sincere gratitude for her adventurous physician.

After my father had consulted Lord Trimblestone, he took me to Drogheda school, of which Doctor Norris was the master, and which was then the best in Ireland. For a few weeks I was there as ridiculous for my English accent, as I had been at Warwick for my Irish brogue. I soon, however, learnt to imitate my companions sufficiently to avoid their ridicule; but I never lost the English pronunciation, to which I was always accustomed at my father's house. When I say that I never lost the English pronunciation, I mean that I never lost it, so that I could not readily resume it at pleasure. After I had at any time resided in England for three months, I returned to the English habits of my early years. Travelling in stage coaches, where people attend much to the idiom of their fellow travellers, I never was taken for an Irishman, though from some organic peculiarity, I believe, in my articulation, or nonarticulation of the letter r, I have frequently been thought to be a native of Cumberland, and sometimes I have been mistaken for a German.

Parents should be very careful about the habits, which their children acquire of pronunciation and idiom; these are scarcely ever to be eradicated in after-life.

The class, in which I was placed at Drogheda school, happened to be composed of the dullest boys under the care of Doctor Norris.—One or two trivial schoolboy anecdotes will be pardoned, as they shew the good humour, and the talents for governing a school, which my excellent master possessed.

We were forbidden to go into a certain street, which was near our play-ground; but this order was as it were by common consent forgotten, or set at nought by most of the boys, upon the following occasion.

An old foreign refugee, a confectioner of the name of Pilioli, had lived in this lane, and was well known to the boys, to whom he used to sell tarts:—he was a merry jesting fellow, and a great favourite with us. Pilioli died suddenly; his funeral was to pass through the forbidden lane, and most of the boys ran out thither to see it.

When we came into school after dinner, Doctor Norris entered with such a frown upon his brow, as quelled the stoutest heart among the culprits. I was not of the number. We stood up to say our lessons; our class was in *Cordery*, and this day's lesson began with " *O Filioli*,"—which, fortunately, the boy who was to begin could not construe. Another and another were equally embarrassed:—at last, when it came to my turn, and when the Doctor sternly reiterated, " *O Filioli*," I looked at him with propitiatory humility, and in a supplicating tone exclaimed "*O Pilioli.*" The Doctor smiled, and pardoned the delinquents.

After I had once reached the head of my class, I kept my place, and then found, that, after I had said my lessons, I had a great deal of weary time on my hands, while the other boys were getting through theirs. It was forbidden to read any thing but our lessons in school-hours. I grew so much tired of having nothing to do, that I constructed a kind of fortress behind a high reading-desk, in which I for some time enjoyed without detection or animadversion the pleasure of amusing myself with English books. I recollect that the book I was reading was Pope's Iliad, which interested me much, when one day, Doctor Norris espied my head behind the reading desk, and I heard him ask, "What is all that?" The boys answered, "It is only Edgeworth's *Cobby-house,* Sir, as he calls it."

"Edgeworth's *what?*"—There was no jesting with the Doctor upon any infringement of the laws.

"Don't you know, Sir, that it is not permitted to read any English books in school-hours?"

I pleaded, that I had said my lesson a considerable time, and that I had nothing to do.

" Why not get your lesson for to-morrow, Sir?"

" I have it, Sir."

" Well, Sir, for the next day?"

" I have it, Sir. — I have my lessons, Sir, for the whole week."

" Stand up this moment then, and say them, Sir, under penalty, that if you miss a word you shall be flogged."

I stood up, and said my lessons for a week, without missing a word. Doctor Norris gave a nod of benevolent approbation, and decreed, that I should thenceforward have permission to read in school-hours, in my *Cobby-house*, whatever English books I chose.

I have all my life felt gratitude to Dr. Norris for the judicious kindness, with which he treated me during the whole time I was at his school : he encouraged me by his approbation, and thus softened the hardship of drudging on with class-fellows, who happened to be dunces. I must except two of my school-fellows, who were remarkably clever boys, and good scholars, the two sons of Chief Baron Foster; John, the eldest, who became afterwards the justly celebrated Speaker of the Irish House of Commons; and William, who became successively Bishop of Kilmore, and Bishop of Clogher. The friendship formed with them at school has lasted through our lives.

Much more than by my scholarship I was distinguished among my companions by my activity in jumping, vaulting, and in every kind of bodily exercise. During vacations from Drogheda school, I was invited to Collon, by Chief Baron Foster, with whose sons I hunted desperately. The eldest, Mr. John Foster, was the best rider that I have ever known. Upon a little horse, that had, I believe from his former possessor, acquired the cognomen of *Beggarman*, I contrived to keep close to the heels of Foster's excellent hunter, often to the admiration of a numerous company of sportsmen.

My father was not fond of hounds or of hunting, so that I had no opportunity of pursuing this royal amusement when at home; I therefore followed shooting, and became so gloriously expert, that I was able to kill eighty per cent at snipe shooting.

When I was about fourteen years old, upon some trifling occasion, I thought myself illtreated at Drogheda, and I prevailed upon my father to remove me to a school in his neighbourhood, at Longford. The name of my new master was Hynes. He was a man of mild manners, a good scholar, and well acquainted with English literature. He attended to me carefully, and I applied with diligence, so that in the course of less than two years I was prepared to enter the University of Dublin.

But, before I went to the University, my attention was suddenly turned from my studies by my eldest sister's marriage. She married Francis Fox, Esq., of Fox Hall, in the county of Longford, a gentleman of good family and good fortune, whose estate being within a few miles of Edgeworth-Town, the families had frequent intercourse. Balls, carousings, and festivities of all kinds, followed my sister's marriage. In these I joined with transports of delight, beyond even what might

have been expected from a boy of my great
vivacity of temper, and personal activity.
Every morning I was following the hounds
with my new brother-in-law, and the foremost in every desperate exploit of the
chase. Every night I was the most incessant, unwearied dancer at the ball. How
human nature, even the nature of a schoolboy, went through all that I did at this
time, I know not. For three nights successively I was never in bed: nor was I
content with all the huntings and dancings
which I have described; but at every interval, when others allowed themselves some
repose, or acknowledged themselvese x-
hausted by fatigue, I was still working off
my superabundant spirit of animation, and
amazing my companions by some extraordinary display of activity. Of several of
these I have in my later years been re-
minded by some of my surviving contemporaries, who have assured me, that they
were eye-witnesses of feats of boyish agility, which I have not only totally forgotten, but can now scarcely believe.

Other circumstances, which happened

about the same time, are more clear in my recollection. My favourite partner among the young ladies at these wedding dances was the daughter of the curate from whom I learned my Accidence.

One night after the dancing had ceased, the young people retired to what was then called a *raking pot of tea*. A description of this Hibernian amusement I have given in another place. It is here sufficient to say, that it is a potation of strong tea, taken at an early hour in the morning, to refresh the spirits of those who have sat up all night. We were all very young and gay, and it was proposed by one of my companions, who had put a white cloak round his shoulders to represent a surplice, that he should marry me to the lady with whom I had danced.

The key of the door served for a ring, and a few words of the ceremony, with much laughter and playfulness, were gabbled over. My father heard of this mock-marriage, and it excited great alarm in his mind. He was induced by his paternal fears to treat the matter too seriously, and he instigated a suit of *jactitation of marriage* in the ecclesiastical court, to annul these imaginary nuptials. The truth was apparent to every body who knew us. No suspicion even was entertained of the young lady's having any design on my heart, or of my having obtained any influence in hers. All the publicity that was given to this childish affair was fortunately of no disadvantage to her; on the contrary, it brought her into notice among persons with whom she might not otherwise have been acquainted, and she was afterwards suitably married in her own neighbourhood. It was before I was sixteen, that I was thus married and divorced. I say *married,* because in the proceedings in this strange suit it was necessary to shew, that a marriage had been solemnized, or else there could have been no divorce.

About this period, much of my spare time was spent at Pakenham Hall, the seat of Lord Longford, the grandfather of the present Earl. He was my father's nephew, and a man of superior abilities and politeness. His lady had also considerable talents, wit,

humour, and a taste for literature, uncommon for women in her day She saw into my
character, or rather, she saw what it might
be made. Field sports then appeared to be
my ruling passion: instead of thwarting my
love for them, she let me shoot till I was
tired; but she gave me the key of the
library, where, as she expected, I soon
passed whole days devouring its contents.
This and other concurring circumstances
extinguished my passion for field sports
before I was eighteen. It never afterwards
revived.

At a great entertainment which was
given at Pakenham Hall, when I was
about fourteen, besides music and dancing,
there was held a Faro bank, at which
the principal gentry played with much
eagerness at no very low rate. Lord
Longford called me aside, and, putting five
guineas into my hand, desired me to try
my fortune. In the course of the evening
I won, what appeared to me, a large sum,
nearly a hundred guineas. The next evening he asked, whether I would again risk
my winning; I readily complied, and

when I was reduced to a single guinea, he offered to lend me whatever I wanted; but I declined this offer, rose from the table, and continued to look on during the rest of the evening. The next day Lord Longford told me, that he had induced me to play, to obtain an insight into my character. " I observed," said he, " that you were never too eager, or too indifferent; that you were not elated when you won, and that you kept your temper when a rapid run of ill-luck reduced you to poverty ; I therefore congratulate you upon your being in all probability exempt from the vice of gaming."

Whether the prophecy, as it frequently happens, became the cause of its own accomplishment, I cannot determine ; but it is certain, that in my subsequent life I never felt an inclination for cards, dice, or lotteries, even when the stake was inconsiderable. By this turn of mind I saved a great deal of time, that is commonly thrown away in weak compliance with the habits of others ; nor did I give offence to those of my companions, who liked play, because they could easily supply my place from the swarm of idlers, who crowd every fashionable assembly.

Immediately after my farcical marriage, and more farcical divorce, I entered Trinity College Dublin, 26th April, 1760. My tutor was the Rev. Patrick Palmer, a gentlemanlike and worthy man; but it was not the fashion in those days to plague fellow-commoners with lectures. My class-fellows, except William Foster my competitor, gave me so little motive for emulation, that I did not trouble myself much with study. In competition with him I was obliged to exert myself strenuously. After a hard fought examination, he obtained from me the premium, which he generously acknowledged to be my right. At the next public examination, I was audaciously and shamefully careless, I went into the hall to translate six books of Homer, of the greatest part of which I had never read one word. A stupid young man succeeded against me, though I certainly answered better than he did; but the examiner, the celebrated Dr. Duigenan, suspecting from my manner, that I had not taken much previous pains,

plainly asked me, how often I had read
these books of Homer. I told him " never."
" Then Sir," said he, " though you have
answered better than your antagonist, I
will not give you the premium, which is
intended as a reward for diligence, and not
as an encouragement for idleness and pre-
sumption."

I wish to pass over my residence at
Dublin College. I was not seventeen. I
was supposed to have some talents, which,
among my associates, was a sufficient apology for my total neglect of study ; and I
passed my time in dissipation of every
kind. It was, however, but for six months,
the only time in my life that I ever spent in
such a disgraceful manner. Young as I
was, I became thoroughly disgusted. I
was sensible of the dangers which I had
incurred, and capable of rejoicing at my
escape. The vigour of my constitution
endured the want of rest and sleep, and
the strength of my head enabled me to re-
sist the effects of those copious potations,
which were in fashion among my companions. My want of taste for the joys of
intoxication, prevented me from continu-

ing the habit, when it was no longer the fashion ; so that I have passed some time at two universities, and have been concerned in conducting four or five contested elections, without ever having been intoxicated in my life.

CHAPTER IV.

1761.]

My father prudently removed me from Dublin to Oxford. I entered Corpus Christi as gentleman-commoner on the 10th of October, 1761. My father preferred Oxford to Cambridge, because he had an old friend, who resided near Oxford; a gentleman who had been bred for the bar, had been with him at the Temple, and upon whose assistance he could depend in the conduct of my studies. This friend was Mr. Elers. To my father's letter, asking his permission to introduce me to his house, Mr. Elers replied, that he should be very glad to be of any service to the son of his old friend; but that, considering the disposition of which I had been described, he thought it right to represent, that he had " several daughters grown and growing up, who, as the world said, were pretty girls; but to whom he could not give fortunes, that could make them suitable matches for Mr. Edgeworth's son."

This letter did not deter my father from his purpose, but probably decided him to put me under the care of such a discreet and honorable friend. My mother was going to Bath at this time for her health, and as soon as he had settled her and his family there, he took me with him to Black Bourton, within fourteen miles of Oxford. Black Bourton, anciently one of the seats of the Hungerford family, had by marriage become the property of my father's friend, Mr. Elers, whose residence it was at this time. As the Elers family became afterwards nearly connected with me, I think it necessary to my history, to give some account of theirs; particularly as there is something uncommon in the rise and fall of their fortunes.

Paul Elers, Esq., was descended from a German family of some opulence. How or why the family of Elerses came into England I know not: but I know, that some of them were favourites of the Elector of Mentz, of whom they had several pictures; I remember one in particular, that had been dismantled of some diamonds, which, from their setting, were probably of great value. The family, declining in England, went over to Ireland, where they engaged in business, and were so successful as to breed up their daughters handsomely, and to maintain their son at the Temple. This son, Paul Elers, was my father's friend. Mr. Elers had applied with such diligence to his studies, that he soon came into reputable business at the English bar. While at the Temple, he not only formed a friendship with my father, but with several other gentlemen of good conduct, and creditable connexions. Among others he was intimate with Mr. Grosvenor, a man of high birth and impaired fortunes. Mr. Grosvenor had played deeply, but it was in the *best* company; and in this society he was singled out by Mr. Hungerford, who had an only daughter reputed to be heiress to a great estate. Mr. Hungerford was pleased with Mr. Grosvenor's manners, and expressed a wish, that he should become his son-in-law, at the same time honorably telling him, that there were some difficulties in making out a title to the estate. A

lawyer of abilities, and fit to be trusted
with family secrets, was consequently to
be employed, to look over the family papers, and to draw such deeds and settle-
ments as were necessary. Paul Elers was
a man of much legal ability and knowledge, of a sober temper, and of a character to be relied on. To him Mr. Grosvenor applied, and partly from friendship,
and partly from the prospect of a handsome
remuneration, he prevailed on Mr. Elers
to accompany him to Black Bourton.
Here Mr. Elers found a family anxious to
know the real situation of their affairs, and
to dispose of an only daughter, who had
never seen the gentleman whom her friends
had chosen for her husband. Some legal
formalities were necessary to secure the
estate to Mr. Grosvenor, who was to possess the whole of it as the fortune of Miss
Hungerford. This lady had not much
beauty, grace, or dignity; she was a plump,
good natured, unfashioned girl, with little
knowledge of any sort, and with no accomplishments. Mr. Grosvenor was not smitten with his intended bride. A fortnight
wore away in turning over parchments,

making out rent rolls, and preparing the terrestrial possessions to be had and held along with his angel. Mr. Grosvenor grew melancholy, and one fair morning expressed his dissatisfaction to Mr. Elers. " The girl is a sad incumbrance on the estate," said he. His friend was of a different opinion, and spoke of Miss Hungerford in terms that astonished Mr. Grosvenor. " A thought," said Grosvenor, " has just struck me ; suppose you were to take the whole bargain off my hands ?" " Most willingly," replied Elers, " if it were possible ; but I fear that neither Mr. Hungerford nor his lady would ever hear of such a proposal." Mr. Grosvenor saw no impossibility. He openly and honorably told his mind to the father and mother. The father liked Mr. Elers, the mother was not so well disposed towards him, but yielded to her husband's arguments ; while the young lady, like Virgil's Lavinia, submitted with blushes, and with becoming filial duty, to the wishes of her parents. Mr. Grosvenor returned with a light heart to London, delighted at his escape; and at having made the fortune of his friend, for such was in all appearance the case. Young Elers had nothing but his profession. He married the heiress, and by his legal skill he was safely secured in the possession of her fortune, an estate of eight hundred pounds a year, highly improveable, well wooded, and within a ring fence. To a young man, without fortune or connexion, such a match as this promised the means of living in ease, if not in affluence; for such an estate, seventy or eighty years ago, was equal to two thousand a year in these days.

But Mr. Elers, by his marriage and new connexions, was at once taken out of the line of life for which he had been educated, and to which he was suited by his talents and early habits. By his application and good conduct, both as a man and as a lawyer, he was coming fast into business, and he had before his marriage a fair prospect of rising to the foremost ranks of his profession. He now became a country gentleman, without connexions, except those of his wife, and without name or influence. He gave up his profession to please his father-in-law, to whom he felt

gratitude that did him honor; but besides
he had considerable expectancies from Mr.
Hungerford, and was bound to him by
prudence as well as gratitude. That industry, which had carried Mr. Elers through
the irksome study of his profession, entirely
failed him in his new situation. He knew
nothing of country business; he had no
taste for field sports, or for the conversation of the neighbouring squires. He had
not acquired the habit of committing his
thoughts to writing, which prevented him
from making any practical use as an author of the stores, which he had laid up in
his capacious memory. In short he had
no object in view, to excite his ambition.
Having no interest in the common routine
of a country life, he had little to do, and
that little he neglected. The family into
which he married was proud, and when
an heir to the family was born, no expense
was spared to celebrate the important
event, and as Mrs. Elers had in perfection one essential quality of a wife, before her husband could look about him,
she had celebrated two or three such fes-
tivals. The lady, had she been ever so

well versed in family economy, could not, during such an incessant production of children, have been of much service in managing the family. Beside the servants necessary in a gentleman's family, there were four or five nurses to be maintained, humored, and kept from breaking the peace. Now Mrs. Elers did not possess any one talent necessary for governing a family, except good humor; and this quality in her arose in some degree from weakness, and from hatred of trouble. A very old steward of the Hungerford family managed all the business of the estate; a great part of which business consisted in choosing, felling, and cutting up wood for fuel. This poor little man, eighty years of age, used to be seen in the depth of winter, upon a little grey horse with shaggy hair and a long flaxen mane and tail, riding about the grounds, and seeming to conduct a number of labourers, who did precisely what they pleased. The value of the timber cut down for firing was more than equal to the price of coals sufficient for the house, and the expense of making it up for use was still greater. Every part of the domestic expenditure was carried on in this manner, so that in a few years after the death of his father-in-law, Mr. Elers found himself in distress, without having been guilty of the slightest extravagance. About twelve years after his marriage, he made an effort to increase his income by chamber-practice, as a lawyer, in the country. In a short time business poured in upon him beyond his most sanguine expectations; and he had again a fair prospect of acquiring some provision for his family. At this time the famous Oxford election took place, and Mr. Elers was looked to by the Marlborough interest, as a proper person to conduct the legal proceedings on that memorable occasion. Mr. Charles Jenkinson was at the same time employed by the same party, and they were considered as candidates of equal pretensions in point of abilities for the favor of those whom they supported. What Mr. Jenkinson's connexions in that country might be, I know not; but Mr. Elers lived among people, who happened to be opposed to the interest which he now supported. His legal acuteness, and, to speak impartially, his readiness to go all lengths, made him extremely obnoxious to those, who had been formerly his best friends and most lucrative clients. How far his complaisance led him I could never distinctly ascertain; but I have heard from both parties, that he was a *useful* and a zealous partizan. After the contest was over, Mr. Elers returned home; and as he was too indolent to pay continual court at Blenheim, or at the houses of his other great friends, he soon lost the interest, which he had made among them. Mr. Jenkinson, on the contrary, never quitted the hold which he had gained, and having constant intercourse with men in power, and being a person of family and address, he advanced steadily in the career of ambition to the height of ministerial eminence.

From the time of the Oxford election, the efforts that Mr. Elers made to advance himself consisted in desultory visits to the houses of the great, and in writing political letters. His letters, though sagacious and pertinent, yet, from his living at a distance from the capital, generally came too late to be of any service. His business as a lawyer forsook him, he had become unpopular, his family rapidly increased, the old steward doated, Mr. Elers left every thing to his wife, and Mrs. Elers left every thing to her servants. Things were in this situation at Black-Bourton, when I was introduced to the family by my father. He had personally known little of Mr. Elers, since their first friendship was formed at the Temple; but judging from his letters, my father considered him as the same man of active mind and talents, and with the same habits for business, which he had then appeared to possess. It was, therefore, naturally a great object with him, to place me on my first going to Oxford under the care of a person whom he so much esteemed, and of whose abilities he had such a high opinion. The family at Black-Bourton at this time consisted of Mrs. Elers, her mother Mrs. Hungerford, and four grown up young ladies, besides several children. The eldest son, an officer, was absent The young ladies, though far from being beauties, were handsome; and though destitute of accomplishments, they were notwithstanding agreeable, from an air of youth and simplicity, and from unaffected good nature and gaiety. The person who struck me most at my introduction to this family group was Mrs. Hungerford. She was near eighty, tall and majestic, with eyes that still retained uncommon lustre. She was not able to rise from her chair without the assistance of one of her grand-daughters; but when she had risen, and stood leaning on her tortoise-shell cane, she received my father, as the friend of the family, with so much politeness, and with so much grace, as to eclipse all the young people by whom she was surrounded. Mrs. Hungerford was a Blake, connected with the Norfolk family. She had formerly been the wife of Sir Alexander Kennedy, whom Mr. Hungerford killed in a duel in Blenheim Park. Why she dropped her title in marrying Mr. Hungerford I know not, nor can I tell how he persuaded the beautiful widow to marry him after he had killed her husband. Mr. Hungerford brought her into the retirement of Black-Bourton, the ancient seat of his family, an excellent but antiquated house, with casement windows, divided by stone frame-work, the principal rooms wainscoted with

oak, of which the antiquity might be guessed from the varnish it had acquired from
time. In the large hall were hung spears,
and hunting tackle, and armour, and trophies of war and of the chase, and a portrait, not of exquisite painting, of the gallant Sir Edward Hungerford. This portrait
had been removed hither from Farley
Castle, the principal seat of the family.
In the history of Mrs. Hungerford there
was something mysterious, which was not,
as I perceived, known to the younger
part of the family. I made no inquiries
from Mr. Elers, but I observed, that she
was for a certain time in the day invisible.
She had an apartment to herself above
stairs, containing three or four rooms; when
she was below stairs, we used to make a
short way from one side of the house to
the other, through her rooms, which occupied nearly one side of a quadrangle,
of which the house consisted. One day,
forgetting that she was in her room, and
her door by accident not having been
locked, I suddenly entered: I saw her
kneeling before a crucifix, which was
placed upon her toilette; her beautiful eyes

streaming with tears, and cast up to Heaven
with the most fervent devotion; her silver
locks flowing down her shoulders; the remains of exquisite beauty, grace, and dignity, in her whole figure. I had not, till
I saw her at these her private devotions,
known that she was a catholic; nor had I,
till I saw her tears of contrition, any reason
to suppose that she thought herself a penitent. The scene struck me, young as I
was, and more gay than young—her tears
seemed to comfort, not to depress her—
and for the first time since my childhood
I was convinced, that the consolations of
religion are fully equal to its terrors. She
was so much in earnest, that she did not
perceive me; and I fortunately had time
to withdraw without having disturbed her
devotions.

But to pursue my own history: I received an unlimited invitation to Black-
Bourton, and soon became one of the family. I laughed, and talked, and sang with
the ladies, and read Cicero and Longinus
with their father, who, notwithstanding
my youth, and my propensity to female
society, filled many of my hours with

agreeable conversation. Having entered Corpus Christi College, Oxford, I applied assiduously not only to my studies, under my excellent tutor Mr. Russell*, but also to the perusal of the best English writers, both in prose and verse. Scarcely a day passed without my having added to my stock of knowledge some new fact or idea; and I remember with satisfaction, the pleasure I then felt, from the consciousness of intellectual improvement.

I had the good fortune to make acquaintance with the young men, the most distinguished at Corpus Christi for application, abilities, and good conduct. When I mention Sir James M‘Donald†, and those with whom he lived, as my companions, I need add nothing more.

My acquaintance with Sir James commenced at the fencing school of Paniotti, a native of one of the Greek islands, a fine old Grecian, full of sentiments of

* Father of the gentleman who is now at the head of the Charter House.

† See Lettres de Madame du Deffand, and Tweddell's Remains.

honor and courage, and of a most independent spirit.

Mr. L., a young gentleman of a noble
family and of abilities, but of overbearing
manners, was our fellow pupil under Paniotti. At the same school we met a young
man of small fortune, and in a subordinate
situation at Maudlin. He fenced in a regular way, and much better than Mr. L.,
who, in revenge, would sometimes take a
stiff foil that our master used for parrying,
and pretending to fence, would thrust it
with great violence against his antagonist. The young man submitted for some
time to this foul play, but at last he ap-
pealed to Paniotti, and to such of his
pupils as were present. Paniotti, though
he had expectancies from the patronage of
the father of his nobly born pupil, yet
without hesitation condemned his conduct.

One day, in defiance of L.'s bullying
pride, I proposed to fence with him, armed
as he was with this unbending foil, on condition that he should not thrust at my face;
but at the very first opportunity he drove
the foil into my mouth. I went to the door,

broke off the buttons of two foils, turned
the key in the lock, and offered one of
these extemporaneous swords to my antagonist, who very prudently declined the
invitation.

This person afterwards shewed through
life an unprincipled and cowardly disposition. The young man, who had at first
borne with him with so much temper, distinguished himself in afterlife in the army.
I mention the circumstance in which I
was concerned, because I believe it contributed to my being well received at first
among my fellow students at Oxford. I
remember with gratitude, that I was liked
by them, and I recollect with pleasure the
delightful and profitable hours I passed at
that University during three years of my life.

Doctor Randolph was at that time
president of Corpus Christi College. With
great learning, and many excellent qualities, he had some singularities, which produced nothing more injurious from his
friends than a smile. He had the habit of
muttering upon the most trivial occasions,
" *Mors omnibus communis.*" One day his

horse stumbled upon Maudlin bridge, and the resigned president let his bridle go, and drawing up the waistband of his breeches as he sat bolt upright, he exclaimed before a crowded audience, " *Mors omnibus communis !*" The same simplicity of character appeared in various instances, and it was mixed with a mildness of temper, that made him generally beloved by the young students. The worthy Doctor was indulgent to us all, but to me in particular upon one occasion, where I fear that I tried his temper more than I ought to have done. The gentlemen-commoners were not obliged to attend early chapel on any days but Sunday and Thursday; I had been too frequently absent, and the president was determined to rebuke me before my companions. " Sir," said he to me as we came out of chapel one Sunday, " You *never* attend Thursday prayers." " I do *sometimes*, Sir," I replied. " I did not see you here last Thursday. And, Sir," cried the president, rising into anger, " I will have nobody in my college," (ejaculating a certain customary guttural noise, some-

thing between a cough and the sound of a postman's horn,) " Sir, I will have nobody in my college that does not attend chapel. I did not see you at chapel last Thursday."
" Mr. President," said I, with a most profound reverence, " it was impossible that you should see me, for you were not there yourself."

Instead of being more exasperated by my answer, the anger of the good old man fell immediately. He recollected and instantly acknowledged, that he had not been in chapel on that day. It was the only Thursday on which he had been absent for three years. Turning to me with great suavity, he invited me to drink tea that evening with him and his daughter. This indulgent president's good humor made more salutary impression on the young men he governed, than has been ever effected by the morose manners of any unrelenting disciplinarian.

During the assizes at Oxford, the gownsmen are or were permitted to seat themselves in the courts. In most country courts there is a considerable share of noise and confusion; but at Oxford the din and

interruption were beyond any thing I have
ever witnessed; the young men were not
in the least solicitous to preserve decorum,
and the judges were unwilling to be severe
upon the students. A man was tried for
some felony, the judge had charged the
jury, and called on the foreman, who
seemed to be a decent farmer, for a verdict. While the judge turned his head
aside to speak to somebody, the foreman of
the jury, who had not heard the evidence
or the judge's charge, asked me, who was
behind him, and whom he had observed
to be attentive to the trial, what verdict
he should give. Struck with the injustice
and illegality of this procedure, I stood up
and addressed the judges Wills and Smith.
" My Lords," said I—" Sit down, Sir,"
said the judge.—" My Lord, I request to
be heard for one moment."—The judge
grew angry.—" Sir, your gown shall not
protect you, I must punish you if you per-
sist."—By this time the eyes of the whole
court were turned upon me; but feeling
that I was in the right I persevered. " My
Lord, I must lay a circumstance before
you which has just happened." The judge

still imagining that I had some complaint to make relative to myself, ordered the sheriff to remove me.—" My Lord, you will commit me if you think proper, but in the mean time I must declare, that the foreman of this jury is going to deliver an illegal verdict, for he has not heard the evidence, and he has asked me what verdict he ought to give."

The judge from the bench made me an apology for his hastiness, and added a few words of strong approbation. This was of use to me, by tending to increase my self-possession in public, and my desire to take an active part in favor of justice.

During vacations, I went to Bath. There I was in imminent danger of being made a coxcomb, by the notice that was taken of me for my dancing. Fortunately, my success in that important art was so complete, as to wear out, before I was twenty, all ambition or vanity upon the subject. I was so much praised for this trifling accomplishment, so much invited as a good dancer, and taken so little notice of for any thing else, that, disgusted and ashamed of myself, I soon began to avoid exhibiting my saltatory talents, and I seldom danced except when it was necessary to make up a set, or to gain an opportunity of conversing with some lady who was agreeable to me.

Bath was at this time filled with the best company. I had now an opportunity of seeing something of what is called *the world*, and of making some observations on characters and manners.

I remember that I was particularly struck with the appearance of the then Duke of Devonshire. He had retired from the court in disgust, and the chagrin visible in his countenance made me early perceive, that the smiles or frowns of princes have more power over the happiness of some human beings, than those who are at a distance from sovereigns can conceive. I saw at the same time ambition on a smaller scale gratified, yet exposed to a certain degree of ridicule; I saw Beau Nash, the subject of a well known epigram, the popular monarch of Bath, whose willing subjects paid him the most implicit and cheerful obedience. Nearly at the same time I saw the wit, who had reigned despotically for above half a century over the world of fashion. I saw

the remains of the celebrated Lord Chesterfield. I looked in vain for that fire,
which we expect to see in the eye of a man
of wit and genius. He was obviously unhappy, and a melancholy spectacle. The
wise and good Lord Littleton was at Bath
during this season. Lord Huntingdon, who
was just returned from his Spanish embassy,
also came there to meet, as it was supposed, the Duke de Nivernois, who made
his appearance for a few days in the rooms.
The Duc de Nivernois' very small hat,
with a most splendid diamond button, attracted the beaus and belles more than his
Excellency's person, which was low, slight,
and not remarkable for beauty. Among
other witticisms of the day, it was said, that
the size of his hat was diminished by the
loss of the Canada fur trade—a loss to the
French, which occurred about this period.

Among all these men of rank and celebrity, Quin, the actor, was not the personage least distinguished in the pump
room at Bath; and at the tavern, " *the
Three Tuns*," he was " without a rival
and without a judge." Another gentle-

man, famous in another species of epicu-
rism, made no inconsiderable figure among
the fair sex. Mr. Medlicott, of gallant
memory. He was my first cousin, and of
course I was acquainted with him. My
father, notwithstanding he was a grave
man, and of strict morality and piety, felt
some fondness for his profligate nephew;
but he had the prudence to put me on my
guard against the danger of his society.
In fact, the dissolute conduct, and more
dissolute conversation of my cousin, served
rather to disgust than to allure me. Upon
the whole this season at Bath was of use
to me, both as to morals and manners. It
formed my manners, by familiarising me
with the best that were then in fashion; and
it wore away in good company that bashful-
ness, which is frequently converted, by living
in other society, into determined awkward-
ness or impudence. My father, believing
that it would be necessary for my happiness to marry early, prudently introduced
me during this season at Bath to families,
where there were daughters such as he
thought would be suitable matches for me.
I became acquainted with several agreeable

young ladies, with whom as their partner in the ball-room I had opportunities of conversing. I soon perceived, that those who made the best figure in a ball-room were not always qualified to please in conversation; I saw, that beauty and grace were sometimes accompanied by a frivolous character, by disgusting envy, or despicable vanity. All this I had read of in poetry and prose; but there is a wide difference, especially among young people, between what is read or related, and what is actually seen. Books and advice make much more impression in proportion as we grow older. We find by degrees, that those who lived before us have recorded as the result of their experience the very things, that we observe to be true. We do not, therefore, continue as we advance in life, to wait for the conviction of our own individual experience; but we endeavour to profit by the example and remarks of others. My observations, however, on female manners and character, and my good father's prudence, did not act time enough to prevent my precipitation.

Before I went to Bath, one of the young ladies at Black-Bourton had attracted my attention; I had paid my court to her, and I felt myself insensibly entangled so completely, that I could not find any honorable means of extrication. I have not to reproach myself with any deceit, or suppression of the truth. On my return to Black-Bourton, I did not conceal the altered state of my mind; but having engaged the affections of the young lady, I married while I was yet a youth at college. I resolved to meet the disagreeable consequences of such a step with fortitude, and without being dispirited by the loss of the society, to which I had been accustomed. I determined to submit to the displeasure of my father with respectful firmness. My mother, though her hopes of me had always been higher than those of my father, yet softened his anger, by suppressing her own feelings of disappointment; and my kind sister, who was a favourite with my father, used all her influence in my favor. By her tears and supplications she obtained his forgiveness. As I was under age I had married in Scotland; but a few months afterwards, my father had me remarried by license with his consent. I had a son before I was twenty; and I soon afterwards took my wife to Edgeworth-Town, to pass a year with my father and mother. Alas! that excellent mother lived only a few days after our arrival. She saw my wife, but could form no judgment of her character. My mother, therefore, exerted herself no more than just to shew her kindness.

On the morning of the day on which my mother died, she called me to her bedside, and told me with a sort of pleasure, that she felt she should die before night. She expressed the following sentiment—" If there is a state of just retribution in another world I must be happy, for 1 have suffered during the greatest part of my life, and I know, that I did not deserve it by my thoughts or actions."

She then communicated to me with great tenderness such remarks upon my character, as she had formed by long and attentive observation. It was then she

said to me those words, which I have recorded* as the advice of an excellent and
wise mother, given with her dying breath.
—"My son, learn how to say NO."—She
warned me further of an error, into which
from the vivacity of my temper I was most
likely to fall—"Your inventive faculty,"
said she, "will lead you eagerly into new
plans; and you may be dazzled by some
new scheme, before you have finished, or
fairly tried what you had begun.—Resolve
to finish, never procrastinate."

These were the last connected sentences
that she uttered. The advice made a due
and lasting impression upon my mind:
after a long life, I cannot now look back
upon any part of my conduct, in which I
neglected this salutary monition. It is a
favourite opinion of many, and it was in
particular a favorite opinion of my friend
Mr. Day, that people never profit by advice; my experience has taught me to believe otherwise. I have frequently profited by the counsel of my friends, and
have frequently known, that others have

* Preface to Vivian.

profited by that which I have had opportunities of giving.

I must be permitted to say a few words
more of a mother, to whom I owe so much.

I believe I have mentioned, that, a few
hours after my birth, she by some mis-
management lost the use of one arm, and
almost of her left side. She was afterwards
afflicted with the stone, so that she lived
in a continual state of bodily pain; and in
a word, her health was most deplorable.
Yet under all these afflictions she was cheerful, and had the full use of her excellent
understanding. Literature was not the fashion of the times when she was young. My
grandmother, as I have been informed, was
singularly averse to all learning in a lady,
beyond reading the Bible, and being able
to cast up a week's household account.
By what accident my mother acquired an
early and a decided taste for knowledge of
all sorts, I never heard; but her application and perseverance were probably sti-
mulated by the preventive measures, that
my grandmother took to hinder her from
wasting time upon books. My mother told

me, that she frequently excused herself
from going to public places and private
parties, that she might obtain an opportunity of reading some favorite author. Partly
from good sense in her choice, and partly
from her good fortune in meeting with their
works, the best authors were her favor-
ites. And at a time when Stella and Mrs.
Delany were looked up to as persons of a
different class from the ladies, who were
commonly to be met with in the best
circles in Ireland, my mother had stored
her mind with more literature, than she
ever allowed to appear in common conversation. The fruits of this early application
amply repaid her for the pains, which she
had taken to cultivate her mind. When
she was incapacitated by disease from any
other enjoyment, she was enabled to lull
the sense of pain by the charms of literature, and by the course of her own thoughts.
I remember, though I was but a child when
I heard them, various conversations between
her and my father upon points of history,
and about opinions upon religious subjects,
in which it appeared to me, that my mother

generally had the advantage: this I collected
sometimes from comparing such of their arguments as I could in some measure comprehend; and at other times, when third
persons competent to decide were present,
I perceived that they usually inclined to
the side, which my mother adopted. Beside fortitude under real sufferings, exem-
plary piety, an excellent understanding,
and much decision of character, she had
the most generous disposition that I ever
met with; not only that common generosity, which parts with money, or money's
worth, freely, and almost without the right
hand knowing what the left hand doth;
but she had also an entire absence of selfish
consideration. Her own wishes or opinions were never pursued merely because
they were her own; the ease and comfort
of every body about her were necessary for
her well-being. Every distress, as far as
her fortune, or her knowledge, or her wit
or eloquence could reach, was alleviated
or removed; and, above all, she could
forgive, and sometimes even forget injuries.
In her own family, domestic order, decent

economy, and plenty were combined; and to the education of her children her whole mind was bent from every ordinary occupation. She had read every thing that had been written on the subject of education, and preferred with sound judgment the opinions of Locke. To these, with modifications suggested by her own good sense, she steadily adhered; and to the influence of her instructions and authority I owe the happiness of my life.

CHAPTER V.

As it was necessary, that I should levy a fine of a small estate, that had been left to me by an uncle, my father kept me in Ireland for a year after my marriage. During that time I read some law, and more science. To amuse myself, I made, with indifferent tools, and with the assistance of an indifferent turner, a wooden orrery, that represented the motions of the sun, moon, and earth. I was then destitute of books to assist me, but I calculated the wheel-work accurately, and invented a movement, to represent the obliquity of the moon's orbit, and its change, which I afterwards found to be the same as what is usually employed in this sort of machinery. I never passed twelve months with less pleasure or improvement: no person of my family had any taste for the scientific employments in which I was occupied, and my young wife in particular had but little sympathy with my tastes. I felt the inconvenience of an early and hasty marriage; and though I heartily repented my folly, I determined to bear with firmness and temper the evil, which I had brought upon myself. Perhaps pride had some share in my resolution.

In the autumn of 1765 I returned to England, and stopped for a few days at Chester, where my wife's aunt resided. By accident I was invited to see the Microcosm, a mechanical exhibition, which was then frequented by every body at Chester. Beside some frivolous moving pictures, the machine represented various motions of the heavenly bodies with neatness and precision. The movements of the figures, both of men and animals, in the pictures, were highly ingenious. I returned so frequently to examine them, that the person who shewed the exhibition was induced to let me see the internal structure of the whole machinery. In the course of conversation, he mentioned the names of some ingenious gentlemen, whom he had met with at different places where he had exhibited, and among the rest he spoke of Doctor Darwin, whom he had met at Lichfield. He described to me a carriage, which the Doctor had invented. It was so constructed, as to turn in a small compass, without danger of oversetting, and without the incumbrance of a crane-necked perch. I determined to try my skill in coach-making, and to endeavour to obtain similar advantages in a carriage of my own construction. As I had no particular object to engage my attention, I had great pleasure in looking forward to this scheme, as a source of employment and amusement. Had I been present at this time of my life in the House of Commons during an animated debate, the subject of which had been level to my capacity, and to the actual state of my knowledge, it is more than probable, that I should have turned my thoughts and my ambition to parliamentary instead of to scientific pursuits.

One evening, whilst I was at Chester, as I was walking with my wife on the walls, I saw at a distance an officer, with whom I had been formerly acquainted when at college in Dublin. I knew him to be one of those dangerous people, who, when they are drunk, "*run a muck*" at all they meet, without distinguishing friend or foe. I perceived that he was intoxicated, and I observed that he lifted up the bonnet of every lady he met, to examine her face. To avoid the quarrel, which must have necessarily ensued if he had taken this liberty with my wife, and finding that I was too near him to retreat, I instantly led her up to him, and introduced her as the bride of his old friend. This not only averted the disagreeable consequences, which would have probably ensued had we met in another manner, but it also prevented him from insulting any body else whilst we continued together. The most unoffending persons are liable sometimes to the dangers of affronts from the folly and intemperance of others; and it is much wiser to run the risk of offending such persons, by fairly meeting them, than to wait for the attack of their capricious humour. I lately (in 1807) happened to see this gentleman again, and adverted to the circumstance. Age had subdued the petulance of his character, but had not effaced the remembrance of his youthful follies. He acknowledged, that my prudence had made a strong and salutary impression upon him at the time, and had probably been of use to him in preventing similar eccentric sallies.

From Chester I went to Black-Bourton, where I found the family in great distress. Mr. Elers was by the malice of an enemy confined for debt. The decline of the fortunes of this family was not occasioned by any extravagance, but brought on by that indolence of character in the head of the house, which I had remarked in my first acquaintance with him. This was afterwards much increased, and added to it was the habit of castle-building. He was continually forming projects of future aggrandizement, upon the civil hopes which were given to him by letters from courtiers; hopes, which he magnified into positive promises. In reward for his services at the famous Oxford election, nothing was done for him, except that an ensign's commission was presented to his eldest son. Yet, though experience did not justify his hopes from the great, still he hoped on from day to day. I have seen him sit hour after hour in his armchair by the fireside, picking cinders from the hearth, and throwing them into the grate: or with his elbows on his knees he would hang over the embers of a wood fire, and from time to time, as any bright idea struck him, he would utter a short cough or ejaculation, sounding like *Hein! Heing!* A sign of secret self-satisfaction in the schemes he was planning.

But while day after day, and year after year, he went on in this manner, building splendid castles in the air, his ancient and magnificent habitation was falling piecemeal to ruins, and his children, for want of ordinary instruction, were growing up in absolute ignorance. Some of the children learned to write from an itinerant writing master, who fortunately was pleased with the simplicity and goodness of his pupils, so that he had taken pains sufficient to teach them to write and spell tolerably well, and to add pounds, shillings, and pence with facility; beyond this they had learned nothing. Of the meaning of many common words in

their own language they had no discriminating notion, farther than that such an
adjective meant to express good or bad, approbation or disapprobation. Yet, while he
thus neglected his own children, Mr. Elers
was assiduous in assisting my studies, not
only by conversation and advice, but by
laborious extracts from books. During his
confinement I used frequently to visit him,
and to endeavour by my conversation and
attentions, to soften the rigor of his situation. I seldom returned without several
sheets of extracts, which between the intervals of my visits he had made for my
instruction from the books he was reading.
These extracts he enriched with his own
judicious remarks. Half this trouble bestowed upon his young children would
have given them an education far superior
to any they could obtain from the temporary lessons of an itinerant master. I relate these facts, not from any want of gratitude to my excellent friend and kind
father-in-law, but to point out an error
that is too common, the bestowing on some
as a gift what is due to others as a debt.

His taste for literature was to Mr. Elers himself a great support even in his most adverse circumstances. In my frequent visits to him at this time, I always found him resigned and cheerful, enjoying his constant resource in books, and applying himself to modern learning in his old age, with the same assiduity as he had in his youth applied to classical studies.

Mean time Mrs Elers was left to manage as well as she could at Black-Bourton, and to take care of a number of helpless children, some of whom were but seven or eight years old. I staid among them for some months, endeavouring to give to the youngest the first rudiments of education, and trying to conduct the affairs of the family in the best manner in my power. It was at last necessary, that I should leave them, and think of my own establishment.

Before I quit Oxfordshire, I may, though unconnected with my own affairs, mention a remarkable circumstance, that happened in the family of a gentleman in that neighbourhood. Mr. Lenthall (descended from the speaker Lenthall) lived at Burford, within a few miles of Black-Bourton. This gentleman, who was a very good master, had a very good butler. One morning the butler came to his master with a letter in his hand, and rubbing his forehead in that indescribable manner, which is an introduction to something which the person does not well know how to communicate, he told Mr. Lenthall, that he was very sorry to be obliged to quit his service.—" Why, what is the matter, John? has any body offended you? I thought you were as happy as any man could be in your situation?"—"Yes, please your honor, that's not the thing, but I have just got a prize in the lottery of 3,000*l*., and I have all my life had a wish to live for one twelvemonth like a man of two or three thousand a year; and all I ask of your honor is, that, when I have spent the money, you will take me back again into your service."— " That is a promise," said Mr. Lenthall, " which I believe I may safely make, as there is very little probability of your wishing to return to be a butler, after having lived as a gentleman."

Mr. Lenthall was however mistaken.

John spent nearly the amount of his ticket in less than a year. He had previously bought himself a small annuity to provide for his old age; when he had spent all the rest of his money, he actually returned to the service of Mr. Lenthall, and I saw him standing at the sideboard at the time when I was in that country.

When I quitted Black-Bourton, I removed to a house, which I had taken upon the common at Hare Hatch, between Reading and Maidenhead, in Berkshire. I had still several terms to keep, before I could be called to the bar, and in the mean time it was necessary for me to live upon a small allowance from my father. My establishment at Hare Hatch was on a very moderate footing. I kept a phaeton with a pair of poneys; a man who took care of them, and of the garden; one man and two maidservants. By the good economy of my wife we lived comfortably. She superintended the care of the garden, which, under her management, was always productive. I had no farm, or any occupation out of doors. The neighbouring houses on the common were inhabited by wealthy but unostentatious people, who were in general contented to visit each other at tea time in the evening, where a game at cards and conversation entertained us till ten o'clock, when we retired to our respective homes. When I look back to this period of my life, I cannot help observing the change of tastes, which has taken place in my mind since that time. I then played at cards three or four times a week; my amusements have long since been very different: thirty, I may say almost forty, years of my life have since passed without a pack of cards having been opened in my house.

While I resided at Hare Hatch, I was in want of amusement, and had no cultivated society. My workshop, and frequent visits to smiths, and coachmakers, and workmen of various sorts, at Reading, occupied the greatest part of the day. My reading was chiefly scientific. It was not till some time afterwards, that I applied myself to general literature; nor till many years afterwards did I suppose, that I should ever become an author. When I went to London to keep term at the Temple, I became acquainted with my brother-in-law, Captain Elers, who lived much with three elderly ladies, Mrs. Blakes, in Great Russel Street, Bloomsbury. He soon grew attached to me, and, indeed, I found him to be one of the best creatures that ever existed. With him I went frequently to the Mrs. Blakes, who were nearly related to the Elers' family. Here I met with people of rank, and particularly with such as deserve the highest rank in society, those who were intent upon doing good. They were many of them indeed neither young nor fashionable; but they were polite and well informed, eager to shew attention to the old ladies whom they visited, and to entertain them by relating what was passing in the world. The Mrs. Blakes seldom went out themselves.

At this time, the celebrated *Comus* exhibited a variety of scientific deceptions in London. I soon discovered many of his secrets. As it was the fashion to go to see him, his tricks became the general subject of conversation; and I was able to entertain a number of Mrs. Blakes' friends,

who were too old or too indolent to go
to his exhibition. Among the ladies who
visited the Mrs. Blakes was a Miss Dalston,
the famous " Fanny, blooming fair," whom
Lord Chesterfield has celebrated. She
was related both to my wife and to Sir
Francis Delaval by the Blakes. One
evening, when she was of the company
assembled at Mrs. Blakes', after I had
been amusing them with some of Comus's
feats, she told us, that her relation, Sir
Francis Delaval, had also discovered these
secrets, and that he believed himself to be
the only man in England who possessed
them. " I shall, however," said she, " inform him, that I have another cousin as
wise as he is."

This slight circumstance first introduced
me to the acquaintance of Sir Francis Delaval. I beg the reader distinctly to understand, that my acquaintance with Sir
Francis commenced but two or three years
before his death. He invited me to his
house, where, in six weeks, I saw more of
what is called *the world,* than I should probably have seen elsewhere in as many

years. I was about two and twenty. Much of what passed before my eyes was not at first perfectly distinct; but I observed, and by degrees various circumstances, that seemed to me extraordinary, and sometimes unaccountable, arranged themselves so as to become scenes as it were of a real comedy—Comedy, I may say, as to the representation before my eyes, but such as had frequently tragic consequences.

At first our joint exhibition of wonders occupied my attention. After arranging our contrivances in the house in Downing Street, where Sir Francis lived, by pre-concerted confederacy, we had it in our power to execute surprising feats. Company of all sorts crowded our exhibitions. Sir Francis was known to every body; but I, as a stranger, was not suspected of being combined with the archfiend in deceiving the spectators. Feats, physically impossible without such assistance, were performed by seeming magic, and many were seriously alarmed by the prodigies which they witnessed. The ingenuity of some of the contrivances, that were employed in our deceptions, attracted the notice not only of those who sought mere amusement, but of men of letters and science, who came to our exhibitions. This circumstance was highly grateful to Sir Francis, and advantageous to me. I, by these means, became acquainted with many men of eminence, to whom I could not at that period of my life have otherwise obtained familiar access. Among the number were Dr. Knight, of the British Museum ; Dr. Watson ; Mr. Wilson ; Mr. Espinasse, the electrician ; Foote, the author and actor, a man, who, beside his well known humor, possessed a considerable fund of real feeling ; Macklin ; and all the famous actors of the day. They resorted to a constant table, which was open to men of genius and merit, in every department of literature and science. I cannot say, that his guests were always " unelbowed by a player ;" but I can truly assert, that none but those who were an honor to the stage, and who were admitted into the best company at other houses, were received at Sir Francis Delaval's. Macklin was our frequent visitor, as he was consulted as to every thing that was

necessary for the getting up of a play, in
which the late Duke of York was to be
the principal actor. On this occasion I
was requested by Sir Francis, to fit up a
theatre in Petty France, near the gate of
the Park, and no trouble and expense were
spared, to render it suitable to the reception of a royal performer. "The Fair
Penitent" was the chosen piece, and the
parts were cast in the following manner,

Sciolto	Mr. J. Delaval.
Horatio	Sir F. Delaval.
Altamont	Sir J. Wrottesly.
Lothario	The Duke of York.
Calista	Lady Stanhope.
Lavinia	Lady Mexborough.

The play was, as to some parts, extremely
well performed. Calista was admirably
acted by Lady Stanhope, and Horatio by
Sir Francis. Sciolto was very well, and
Lothario was as warm, as hasty, and as
much in love, as the fair Calista could possibly wish. After the piece, Sir Francis
and his friends from the real theatres retired to sup, and to criticise, at the King's
Arms, Covent Garden. It was singular

that Sir Francis, who was the projector of
the scheme, preferred supping with his critical friends to partaking of an entertainment with the Duke of York, and a splendid company. I accompanied Sir Francis
Delaval, and we passed a most agreeable
evening. The company were, in fact, all
performing amusing parts, though they
were off the stage. After we had supped,
Macklin called for a nightcap, and threw
off his wig. This, it was whispered to me,
was a signal of his intention to be entertaining. Plays, playwrights, enunciation,
action, every thing belonging to eloquence
of every species, was discussed. Angelo,
the graceful fencing-master, and Bensley,
the actor, were of the party; Angelo was
consulted by Bensley, on what he ought to
do with his hands while he was speaking.
Angelo told him, that it was impossible to
prescribe what he should always do with
them; but that it was easy to tell him what
should *not* be done—" he should not put
them into his breeches' pockets"—a custom
to which poor Bensley was much addicted.
  Pronunciation was discussed; the faults
in our language in this particular were

copiously enumerated. " For instance," said Macklin, " *Pare* me a *pair* of *pears*." You may take three words out of this sentence, of the same sound, but of different meaning, and I defy any man to pronounce them in such a manner as to discriminate the sounds, or to mark to any ear by his pronunciation the difference between the verb, *to pare*, the noun of number, *a pair*, and the fruit, *pear*. The pompous Bensley undertook that Powel, who was remarkable for a good ear, should do this. Bensley, who mouthed prodigiously whilst he spoke, was put behind a curtain, that the motion of his lips might not assist Powel in judging what meaning he intended to express by each of the words as he pronounced them. One of the company was placed behind the curtain, and to him Bensley was previously to communicate, whether he proposed to pronounce the word denoting the action, the noun of number, or the fruit. Bensley failed so often, and so ridiculously, that he became quite angry, and charged Powel with wilful misapprehension. To defend himself, Powel proposed that Holland should try his skill; but Holland had no better success. During these trials, I concerted by signs with Sir Francis a method of pointing out my meaning, and I offered to try my skill. The audience with difficulty restrained their contempt; but I took my place behind the curtain, and they were soon compelled to acknowledge, that I had a more distinct pronunciation, or that Sir Francis had more accurate hearing, than the rest of the company. Out of twenty experiments, I never failed more than two or three times, and in these I failed on purpose to prevent suspicion. I had made my confederate understand, that when I turned my right foot outward, as it appeared from beneath the curtain, I meant to say *pare*, to cut; when I turned it inward, *pair*, a couple; and when it was straight forward, *pear*, the fruit. We kept our own counsel, and won unmerited applause. Amidst such trifling as this much sound criticism was mixed, which improved my literary taste, and a number of entertaining anecdotes were related, which informed my inexperienced mind with knowledge of the world.

In his youth, Sir Francis Delaval had a
great love of frolic, and now, when he became intimate with me, he related to me
some of the adventures of his early life, a
few of which I may here mention.

Once, when he stood for the Borough of
Andover, an opposition took place, and the
corporation was so closely divided, that it
was nearly a drawn battle between him and
his competitor. One sturdy fellow among
the voters held out against all applica-
tions: he declared, that he would vote for
neither of the contending candidates. Sir
Francis paid him a visit, and with much
address endeavoured to discover some
means of softening him. Sir Francis knew,
that the man was unassailable by plain
bribery; he therefore tried to tempt his
ambition, his love of pleasure, his curiosity,
in short, every passion that he thought
could actuate this obstinate voter. Sir
Francis found, that all the public spectacles
of London were familiar to this man, who
had often gone to town, on purpose to see
various exhibitions. This seemed to have
been his favorite relaxation. After many
attempts, Sir Francis at last discovered,

that this odd mortal had never seen a fire-eater, and that he did not believe the wonderful stories he had heard of fire-eaters; nor could it, he said, be imagined, that any man could vomit smoke, and flame, and fire from his mouth like a volcano. Sir Francis proposed to carry him immediately to town, and to shew him the most accomplished eater of fire that had ever appeared. The wary citizen of Andover suspected some trick, and could by no means be prevailed upon to go up to town. Our staunch candidate, never at a loss for resource, despatched instantly a trusty servant to London, requesting Angelo to come to his assistance. Among his various accomplishments, Angelo possessed the art of fire-eating in the utmost perfection; and though no pecuniary consideration could have induced him to make a display of his talents, in such an art, yet to oblige Sir Francis, to whom all his friends were enthusiastically devoted, Angelo complied. A few hours after he received the request, he thundered into Andover in a chaise and four, express, to eat fire for Sir Francis Delaval's friend! When the obdurate voter saw this gentleman come down, and with such expedition, on purpose to entertain him, he began to yield. But when Angelo filled his mouth with torrents of flame, that burst from his lips and nostrils, and seemed to issue even from his eyes; when these flames changed to various colours, and seemed continually to increase in volume and intensity; our voter was quite melted: he implored Angelo to run no farther hazard; he confessed, " that he did not think the devil himself could cast out such torrents of fire and flame, and that he believed Sir Francis had his Satanic Majesty for his friend, otherwise Sir Francis never could have prevailed upon him to break the vow which he had made not to vote for him.

For this time Sir Francis succeeded in his election; but on the next occasion he found his interest still lower than before in Andover. When he commenced his canvas, he went to the house of the mayor of Andover, who had hitherto been his friend, and with whom he usually lodged. The mayor's lady had also been on his side formerly, but Sir Francis now perceived by her

averted glances, that he had lost her favor.
As he paid her some compliments while
she made tea, the lady scornfully replied,
that "his compliments to her tea were no
more genuine than his tea-canisters." Now
it seems that on the former occasion a
promise had been made to her of a handsome tea-chest with silver canisters, in place
of which she had received only plated canisters. Sir Francis was struck dumb by
this discovery. When he recovered himself, he protested in the most energetic
manner, that this trick had been put upon
him as well as upon her, by the person
whom he had employed to purchase the
tea-chest. He offered to produce his order
to his agent, he pleaded his own character
as a gentleman, and his known habits, not
only of generosity, but of profusion. All
would not do, the enraged mayoress treated
his apologies with disdain, and his pro-
fessions as counterfeit coin. What was to
be done? With the mayor's vote he lost
other voices. The corporation openly declared, that unless some person of wealth,
and consequence, and honor, appeared from
London, and proposed himself candidate,

they would elect a gentleman in the neighbourhood, who had never canvassed the borough, rather than let Sir Francis come in.
Next morning an express arrived early in
Andover, with an eloquent and truly polite
letter from Sir Robert Ladbroke, who
was then father of the city, declaring his
intention to stand candidate for the free
and independent borough of Andover, in-
timating that his gouty state of health required care, and begging the mayor, with
whom he had some acquaintance, to secure
for him a well-aired lodging. Mrs. Mayoress, in high exultation, had a bed prepared for the infirm Sir Robert in her best
bed-chamber; supper was ready at an
early hour, but no Sir Robert appeared.
At length a courier arrived with a letter
excusing his presence that night, but promising that Sir Robert would breakfast
next morning with the mayor. In the
mean time the neighbouring gentleman,
who had been thought of as rival candidate
to Sir Francis Delaval, not finding himself
applied to, and seeing no likelihood of
success, had prudently left home to avoid
being laughed at. The morning came,

the breakfast passed, and the hour of election approached. An express was sent to hurry Sir Robert. The express was detained on the road, and when the writ was to be read, and the books opened, the old member Sir Francis Delaval appeared unopposed on the hustings; his few friends gave their votes, and in default of the expected Sir Robert, who was never forthcoming, Sir Francis was duly elected.

Here ended Sir Francis Delaval's electioneering successes at Andover. His attorney's bill was yet to be discharged. It had been running on for many years, and though large sums had been paid on account, a prodigious balance still remained to be adjusted. The affair came before the King's Bench. Among a variety of exorbitant and monstrous charges there appeared the following article.

" To being thrown out of the window at the George Inn, Andover—to my leg being thereby broken—to surgeon's bill, and loss of time and business—all in the service of Sir F. B. Delaval.—Five hundred pounds."

When this curious *item* came to be explained, it appeared, that the attorney had,

by way of promoting Sir Francis's interest
in the borough, sent cards of invitation to
the officers of a regiment in the town, in
the name of the mayor and corporation,
inviting them to dine and drink His Majesty's health on his brthday. He, at
the same time, wrote a similar invitation
to the mayor and corporation, in the name
of the officers of the regiment. The two
companies met, complimented each other,
eat a good dinner, drank a hearty bottle
of wine to His Majesty's health, and prepared to break up. The commanding
officer of the regiment, being the politest
man in company, made a handsome speech
to Mr. Mayor, thanking him for his hospitable invitation and entertainment. "No,
colonel," replied the mayor, " it is to you
that thanks are due by me and by my
brother aldermen for your generous treat to
us." The colonel replied with as much
warmth as good breeding would allow: the
mayor retorted with downright anger,
swearing that he would not be choused by
the bravest colonel in His Majesty's service.—" Mr. Mayor," said the colonel,
" there is no necessity for displaying any

vulgar passion on this occasion. Permit me to shew you, that I have here your obliging card of invitation."—" Nay, Mr. Colonel, here is no opportunity for bantering, there is your card."

Upon examining the cards, it was observed, that, notwithstanding an attempt to disguise it, both cards were written in the same hand by some person, who had designed to make fools of them all. Every eye of the corporation turned spontaneously upon the attorney, who, of course, attended all public meetings. His impudence suddenly gave way, he faltered and betrayed himself so fully by his confusion, that the colonel, in a fit of summary justice, threw him out of the window. For this Sir Francis Delaval was charged five hundred pounds.—Whether he paid the money or not, I forget.

Some years before I was acquainted with him, Sir Francis, with Foote for his coadjutor, had astonished the town as a conjuror, and had obtained from numbers vast belief in his necromantic powers. This confidence he gained, chiefly by relating to those who consulted him the past events of their lives; thence he easily persuaded them, that he could foretell what would happen to them in future; and this persuasion frequently led to the accomplishment of his prophecies. Foote chose for the scene of their necromancy a large and dark room in an obscure court, I believe in Leicester Fields. The entrance to this room was through a very long, narrow, winding passage, lighted up by a few dim lamps. The conjuror was seated upon a kind of ottoman in the middle of the room, with a huge drum before him, which contained his familiar spirit. He was dressed in the eastern fashion, with an enormous turban, and a long white beard. His assistant held a white wand in his hand, and with a small stick struck the drum from time to time, from which there issued a deep and melancholy sound. His dragoman answered the questions that were asked of him by his visitants, while the conjuror preserved the most dignified silence, only making signs, which his interpreter translated into words. When a question was asked, the visitant was kept at a distance from the drum, from which the oracle seemed to proceed. The former habits, and extensive acquaintance of Sir F. Delaval, and of his associates, who, in fact, were all the men of gallantry of his day, furnished him with innumerable anecdotes of secret intrigues, which were some of them known only to themselves and their paramours. Foote had acquired a considerable knowledge of the gallantries of the city; and the curiosity, which had been awakened and gratified at the west end of the town by the disclosure of certain ridiculous adventures in the city, gave to the conjuror his first celebrity. It was said, that he had revealed secrets that had been buried for years in obscurity. Ladies as well as gentlemen among the fools of quality were soon found, to imitate the dames of the city in idle and pernicious curiosity; and under the sanction of fashion, the delusion spread rapidly through all ranks. Various attempts were made to deceive the conjuror under false names, and by a substitution of persons; but he in general succeeded in detecting these, and his fame stood at one time so high, as to induce persons of *the first considera-*

tion to consult him secretly. His method of obtaining sudden influence over the incredulous was by telling them some small detached circumstances, which had happened to them a short time before, and which they thought could scarcely be known to any body but themselves. This he effected by means of an agent, whom he employed at the door as a porter. This man was acquainted with all the intriguing footmen in London, and whilst he detained the servants of his master's visitants as they entered, he obtained from them various information, which was communicated by his fellow servants through a pipe to the drum of the conjurer. It was said, that in the course of a few weeks, while this delusion lasted, more matches were made, and more intrigues were brought to a conclusion, by Sir Francis Delaval and his associates, than all the meddling old ladies in London could have effected or even suspected in as many months. Among the marriages was that of Lady Nassau Paulet with Sir Francis himself. This was the great object of the whole contrivance. As soon as it was accomplished, the conjurer prudently decamped, before an inquiry too minute could be made into his supernatural powers. Lady Nassau Paulet had a very large fortune, I believe eighty thousand pounds, of all which Sir Francis Delaval became possessed by this marriage. Her ladyship died soon afterwards, and her fortune did not long continue to console her husband for her loss. The whole of the eighty thousand pounds he contrived soon to dissipate.

CHAPTER VI.

————

WHATEVER knowledge of the world Sir Francis Delaval and Foote had acquired, I collected at an easy rate from their conversation. The love of adventure was not quite extinguished in Sir Francis, when I first knew him. It was some time after the death of Lady Nassau Paulet, and Sir Francis was looking out for another wife, and another fortune.

Lady Jacob, the widow of a Sir something Jacob, was then an object of pursuit to the fortunehunting men of fashion, and Sir Francis was of the number. His rivals were mere empty coxcombs. During several tiresome evenings, that I walked the round of Ranelagh in their company, I never heard from them a single sentiment or expression worth repeating. Grimace, and a waterdog shake of the head, supplied the place of conversation. The widow had some cleverness, though it had not been much cultivated by education; and I plainly saw her disgust at the nonsense which she endured. I remember that one of her honorable lovers, after a composing prelude of fashionable fatuity, with a solemn air and complacent smile requested his mistress's opinion upon the propriety of having the candles snuffed.

The lady saw Sir Francis Delaval's superiority to his competitors in abilities and address, but his character for gallantry could not be unnoticed by the wary widow. She laughed at his rivals for attempting to vie with him, but at the same time she told him, that, though he was far superior to any of them in talents and accomplishments, yet she must be sure of his reformation, before she could venture to make him her master. That he must undergo a moral probation. This was a species of trial not much to the taste of Sir Francis, he therefore abandoned the field to his insipid rivals.

The widow, after examining maturely the pretensions of these various suitors, wisely dismissed them all, and married a young Irish captain, whose claims to her favor fairly rested on his sword and his figure.

Sir Francis Delaval was soon afterwards engaged with other objects. He had a universal acquaintance with all the gay and all the gambling world. Lord March, afterwards Duke of Queensberry, Jennison Shaftoe, Lord Eglintoun, Mr. Thynne, Lord Effingham, Colonel Brereton, and numbers, whose names have long since been forgotten, consulted Sir Francis in their schemes at Newmarket; his ingenuity and never failing resources made his acquaintance highly valuable to such gentlemen of the turf-club, as made bets out of the common line of gambling. A coachmaker's journeyman had been taken notice of by Lord March, for his being able to run with a wheel upon the pavement with uncommon speed, which his lordship had ascertained at leisure with his stop-watch. A waiter in Betty's fruit shop, in St. James's Street, was also famous for running. His speed

Lord March minuted, and upon some opportunity he spoke of the coachmaker's running, as if he believed, that the wheel assisted instead of retarding his speed. This brought on discussion, and Lord March offered to lay a large wager, that the coachmaker's journeyman should run with the wheel of his Lordship's carriage, which was at the door, faster than the waiter who was in the room. The bet was taken up to a considerable amount, and the time and place determined. Lord March well knew, that large bets would depend on each side among the frequenters of the *turf;* and that each of the competitors would be engaged to try their speed, that those who backed them might know what they had to depend upon. He, therefore, had the waiter carefully watched, and had his speed ascertained; he also had experiments tried by the journeyman coachmaker. By these means he thought himself almost certain of success, and he and his friends took up as many bets as they could before the day appointed for the race. The gentlemen on the other side

had not been inattentive; and having observed, that the coachmaker always ran
with one particular wheel, which was considerably higher than that with which Lord
March had betted he should run; and
being well-assured by coachmakers, whom
they consulted, that a man could not roll a
small wheel nearly so fast as a large one;
they reckoned upon this circumstance as
decisive in their favor, because the hind
wheel of Lord March's carriage happened
to be uncommonly small. By some means
their hopes in this advantage was discovered, but not till the very day before the
match was to be determined. Lord March
immediately tried the rate of his racer,
with the wheel with which he was actually
to run, and found such an evident difference from that upon which he had depended, as to leave him very little chance
of success. He mentioned his distress to
Sir Francis Delaval, who instantly suggested a remedy. He applied immediately
to friends whom he had in the board of
works, for a number of planks sufficient to
cover a pathway on the course, where the

men were to run. By the help of numbers, with the aid of moonlight, he laid these planks upon blocks, of a height sufficient to raise the nave of his low wheel to the height of that with which the coachmaker had been accustomed to run. The jockey-club allowed the expedient, and Lord March won his wager.

Bets of this sort were in fashion in those days, and one proposal of what was difficult and uncommon led to another. A famous match was at that time pending at Newmarket between two horses, that were in every respect as nearly equal as possible. Lord March, one evening at Ranelagh, expressed his regret to Sir Francis Delaval, that he was not able to attend Newmarket at the next meeting. " I am obliged," said he, " to stay in London ; I shall, however, be at the Turf Coffee-House ; I shall station fleet horses on the road, to bring me the earliest intelligence of the event of the race, and I shall manage my bets accordingly."

I asked at what time in the evening he expected to know who was winner.—He said about nine in the evening. I asserted,

that I should be able to name the winning
horse at four o'clock in the afternoon. Lord
March heard my assertion with so much
incredulity, as to urge me to defend myself; and at length I offered to lay five
hundred pounds, that I would in London
name the winning horse at Newmarket, at
five o'clock in the evening of the day when
the great match in question was to be run.
Sir Francis having looked at me for encouragement, offered to lay five hundred
pounds on my side; Lord Eglintoun did
the same; Shaftoe and somebody else took
up their bets; and the next day we were
to meet at the Turf Coffee-House, to put
our bets in writing. After we went home,
I explained to Sir Francis Delaval the
means that I proposed to use. I had early
been acquainted with Wilkins's " Secret
and swift Messenger;" I had also read in
Hooke's Works of a scheme of this sort, and
I had determined to employ a telegraph
nearly resembling that which I have since
published*. The machinery I knew could
be prepared in a few days.

* In the memoirs of the Royal Irish Academy, and in
Nicholson's Journal for October, 1798, quarto, vol. ii,
p. 320.

Sir Francis immediately perceived the feasibility of my scheme, and indeed its certainty of success. It was summer time, and by employing a sufficient number of persons, we could place our machines so near as to be almost out of the power of the weather. When we all met at the Turf Coffee-House, I offered to double my bet, so did Sir Francis. The gentlemen on the opposite side were willing to accept my offer; but before I would conclude my wager, I thought it fair to state to Lord March, that I did not depend upon the fleetness or strength of horses to carry the desired intelligence, but upon other means, which I had, of being informed in London which horse had actually won at Newmarket, between the time when the race should be concluded and five o'clock in the evening. My opponents thanked me for my candor, reconsidered the matter, and declined the bet. My friends blamed me extremely for giving up such an advantageous speculation. None of them, except Sir Francis, knew the means which I had intended to employ, and he kept them a profound secret, with a view

to use them afterwards for his own purposes. With that energy, which charac-
terised every thing in which he engaged,
he immediately erected, under my directions, an apparatus between his house and
part of Piccadilly; an apparatus, which
was never suspected to be telegraphic. I
also set up a night telegraph between a house
which Sir F. Delaval occupied at Hampstead, and one to which I had access in
Great Russell Street, Bloomsbury. This
nocturnal telegraph answered well, but
was too expensive for common use.

Upon my return home to Hare Hatch,
I tried many experiments on different
modes of telegraphic communication. My
object was to combine secrecy with expedition. For this purpose I intended to
employ windmills, which might be erected
for common economical uses, and which
might at the same time afford easy means
of communication from place to place upon
extraordinary occasions. There is a wind-
mill at Nettlebed, which can be distinctly
seen with a good glass from Assy Hill,
between Maidenhead and Henley, the
highest ground in England, south of the

Trent. With the assistance of Mr. Perrot, of Hare Hatch, I ascertained the practicability of my scheme between these places, which are nearly sixteen miles asunder.

I have had occasion to shew my claim to the revival of this invention in modern times, and in particular to prove, that I had practised telegraphic communication in the year 1767, long before it was ever attempted in France. To establish these truths, I obtained from Mr. Perrot, a Berkshire gentleman, who resided in the neighbourhood of Hare Hatch, and who was witness to my experiments, his testimony to the facts which I have just related. I have his letter; and, before its contents were published in the Memoirs of the Irish Academy for the year 1796, I shewed it to Lord Charlemont, President of the Royal Irish Academy.

During my residence at Hare Hatch, another wager was proposed by me among our acquaintance, the purport of which was, that I undertook to find a man, who should, with the assistance of machinery, walk faster than any other person that could be produced. The machinery which I intended to employ was a huge hollow wheel made very light, within side of which, in a barrel of six feet diameter, a man should walk. Whilst he stepped thirty inches, the circumference of the large wheel, or rather wheels, would revolve five feet on the ground; and as the machine was to roll on planks, and on a plane somewhat inclined, when once the *vis inertiæ* of the machine should be overcome, it would carry on the man within it, as fast as he could possibly walk. I had provided means of regulating the motion, so that the wheel should not run away with its master. I had the wheel made, and when it was so nearly completed as to require but a few hours' work to finish it, I went to London for Lord Effingham, to whom I had promised, that he should be present at the first experiment made with it. But the bulk and extraordinary appearance of my machine had attracted the notice of the country neighbourhood; and taking advantage of my absence, some idle curious persons went to the carpenter I employed, who lived on Hare Hatch common. From him they obtained the great wheel, which had been left by me in his care. It was not finished. I had not yet furnished it with the means of stopping or moderating its motion. A young lad got into it, his companions launched it on a path which led gently down hill towards a very steep chalk-pit. This pit was at such a distance, as to be out of their thoughts, when they set the wheel in motion. On it ran. The lad withinside plied his legs with all his might. The spectators, who at first stood still to behold the operation, were soon alarmed by the shouts of their companion, who perceived his danger. The vehicle became quite ungovernable, the velocity increased as it ran down hill. Fortunately the boy contrived to jump from his rolling prison before it reached the chalk-pit; but the wheel went on with such velocity, as to outstrip its pursuers, and, rolling over the edge of the precipice, it was dashed to pieces.

The next day, when I came to look for my machine, intending to try it upon some planks, which had been laid for it, I found to my no small disappointment, that the object of all my labors and my hopes was lying at the bottom of a chalk-pit, broken into a thousand pieces. I could not at that time afford to construct another wheel of this sort, and I cannot therefore determine what might have been the success of my scheme.

As I am on the subject of carriages, I shall mention a sailing carriage, that I tried on this common. The carriage was light, steady, and ran with amazing velocity. One day, when I was preparing for a sail in it, with my friend and school-fellow, Mr. William Foster, my wheel-boat escaped from its moorings, just as we were going to step on board. With the utmost difficulty I overtook it, and as I saw three or four stage-coaches on the road, and feared that this sailing chariot might frighten their horses, I, at the hazard of my life, got into my carriage while it was under full sail, and then, at a favorable part of the road, I used the means I had of guiding it easily out of the way. But the sense of the mischief which must have ensued, if I had not succeeded in getting into the machine at the proper place, and stopping it at the right moment, was so strong, as to deter me from trying any more experiments on this carriage in such a dangerous place.

Such should never be attempted except on a large common, *at a distance from a high road.* It may not however be amiss to suggest, that upon a long extent of iron rail-way, in an open country, carriages properly constructed might make profitable voyages from time to time with sails instead of horses; for though a constant or regular intercourse could not be thus carried on, yet goods of a certain sort, that are saleable at any time, might be stored till wind and weather were favorable.

When the time came for completing my terms at the Temple, I went again to London, and my intimacy with Sir F. Delaval was renewed. Beside the incidental schemes and amusements which I have mentioned, one great object had long filled his mind. The Duke of York was in love with Sir. Francis Delaval's sister, Lady Stanhope. Her husband, Sir William Stanhope, was dying, and the great object was to keep the duke's flame alive. Every body of abilities about the duke, whom Sir Francis could influence, was engaged in supporting this project.

But the hand of death put a stop to the scheme. The Duke of York, in a tour to Italy, went to some ball in Rome, and, after dancing violently, caught cold in returning by night to his residence, which was at a considerable distance from the place of entertainment: he was seized with a fever, and died. Suspicions of poison arose; but the Prince of Monaco, at whose palace he died, came over to London, and dissipated this surmise.

By the death of the Duke of York, Sir Francis found all his schemes of aggrandisement blasted. Though a man of great strength of mind, and of vivacity that seemed to be untameable, his spirits and health sunk under this disappointment. His friends and physician laughed at his complaints. Of Herculean strength, and, till this period, of uninterrupted health, they could not bring themselves to believe, that a pain in his breast, of which he complained, was of any serious consequence; on the contrary, they treated him as an hypochondriac, whom a generous diet, amusement, and country air, would soon restore. He was ordered, however, to use a steam-bath, which was then in vogue, at Knightsbridge. I went with him there one day, the last I ever saw him! He expressed for me a great deal of kindness and esteem : and then seriously told me he felt, that, notwithstanding his natural strength both of body and mind, and in contradiction of the opinion of all the physicians, he had not long to live. He acknowledged, that his mind was affected as well as his body.

" Let my example," said he, " warn you
" of a fatal error, into which I have fallen,
" and into which you might probably fall,
" if you did not counteract the propensities,
" which might lead you into it. I have
" pursued amusement, or rather frolic, in-
" stead of turning my ingenuity and talents
" to useful purposes. I am sensible," continued he, " that my mind was fit for
" greater things, than any of which I am
" now, or of which I was ever supposed
" to be capable. I am able to speak flu-
" ently in public, and I have perceived, that
" my manner of speaking has always in-

" creased the force of what I have said.
" Upon various useful subjects I am not
" deficient in information; and if I had
" employed half the time and half the pains
" in cultivating serious knowledge, which
" I have wasted in exerting my powers
" upon trifles, instead of making myself
" merely a conspicuous figure at public
" places of amusement, instead of giving
" myself up to gallantry which disgusted
" and disappointed me, instead of dissi-
" pating my fortune and tarnishing my
" character, I should have distinguished
" myself in the senate or the army, I should
" have become a USEFUL member of society,
" and an honour to my family. Remem-
" ber my advice, young man! Pursue what
" is USEFUL to mankind, you will satisfy
" them, and, what is better, you will satisfy
" yourself."

Two mornings afterwards he was found
dead in his bed*. Thus ended Sir Francis

* His friends, perhaps to obviate any suspicion of his
having destroyed himself, had his body opened, and the
physician, who attended, informed me, that his death was
probably occasioned by an unnatural distention of his stomach, which seemed to have lost the power of collapsing.

Blake Delaval. Descended from illustrious ancestors, born with every personal advantage, of a countenance peculiarly prepossessing, tall, strong, athletic, and singularly active, he excelled in every manly exercise, was endowed with courage, and with extraordinary presence of mind; yet all in vain. His parting advice was not thrown away upon me. Indeed I had heard and seen sufficient to convince me, that a life of pleasure is not a life of happiness, and that to the broad gaiety of public festivity there frequently succeeds insupportable ennui in private—ennui, which often drives men to the worst vices merely for emotion and occupation.

This they attributed to his drinking immoderate quantities of water and small beer. He always had a large jug of beer left by his bed-side at night, which was usually empty before morning.—Whether this was cause or effect still remains uncertain.

CHAPTER VII.

———

AFTER Sir Francis Delaval's death, I returned home, and resumed my occupations and my amusements in mechanics. These led me to an acquaintance with Mr. Gainsborough, a Presbyterian clergyman, brother to the celebrated painter of that name. He lived at Henley upon Thames, within a few miles of me. We became intimately acquainted, and I do not think, that I have ever known a man of a more inventive genius.

As many parts of the high land in the neighbourhood of Henley were ill supplied with water, every contrivance, that promised to facilitate the means of raising it, was eagerly adopted. This induced Mr. Gainsborough, to turn his thoughts to this subject. His inventive faculty might have been employed more advantageously; for it must be obvious even to those who are but slightly conversant with mechanics, that no possible application of the power of men or animals can alter their effect in any considerable degree, and that the application of wind is too variable, and of steam commonly too complicated, for domestic purposes. He notwithstanding erected several ingenious hydraulic machines in various parts of the country, which shewed a fertile invention, and in all their parts a sound knowledge of the principles of mechanics. In many instances he gave a large scope to his genius, in obviating local difficulties, and inventing tools to execute his purposes in country places, where he could not enjoy the resources of the capital. He was besides an excellent and most accurate workman, and had he early turned his thoughts to the construction of timepieces for ascertaining the longitude, I make no doubt, that he would have succeeded as well as any man, who could have been his competitor. I believe I took from him hints of some small contrivances, which I have since executed; but were he alive, he would not complain. He was too much my friend, and he was possessed of too much generosity, to suppose that I was either so poor in resources, or so mean in disposition, as to steal from any man.

Amongst other contrivances, I remember to have seen a dial, which shewed time distinctly to one minute, without the assistance of wheelwork or microscopes. For a tide-mill of his invention he obtained a premium of 50*l*. from the Society for the Encouragement of Arts.

One word more in remembrance of this worthy and ingenious gentleman. He planned and executed the handsome road, that goes up the steep hill near Henley. In cutting down this hill, he employed carriages without horses, by means of a large horizontal pully, which enabled the full carriages, as they went down the hill, to draw up the empty ones.

I formerly mentioned, that on my return from Ireland, when I staid a day at Chester, I had heard from the owner of the Microcosm, that Dr. Darwin, of Lichfield, had invented and constructed a carriage upon a new principle. Upon the principle, that, in turning round, it con-

tinued to stand on four points, nearly at
equal distances from each other; whereas
in carriages with a crane-neck, when the
four wheels are *locked* under the perch, the
fore carriage is very unsteady, being supported upon only three points. From this
hint, without having seen the Doctor's
carriage, I constructed a very handsome
phaeton upon this principle. I had it with
me in London. Sir Francis Delaval had
taken a fancy to it, and had ordered another from Monk, the person who had
made mine, and who upon this occasion
first came from the country to settle in
Town. Upon its being approved of by
the Society for the Encouragement of Arts,
both on account of the manner in which
the fore carriage locked, and also on account of a sure and simple method of dis-
engaging horses should they become un-
ruly, I told the society, that I had taken
the hint of the contrivance for preventing
accidents to a carriage in turning, from a
description that had been given me of a
carriage of Dr. Darwin's, of Lichfield; I
wrote an account to the Doctor of the reception, which his scheme had met with

from the Society of Arts &c. The Doctor
wrote me a very civil answer; and though,
as I afterwards found out, he took me for
a coachmaker, he invited me to his house.
An invitation which I accepted in the ensuing summer.

When I arrived at Lichfield, I went to
inquire whether the Doctor was at home.
I was shewn into a room, where I found
Mrs. Darwin. I told her my name. She
said the Doctor expected me, and that he
intended to be at home before night. There
were books and prints in the room, of
which I took occasion to speak. Mrs.
Darwin asked me to drink tea, and I perceived, that I owed to my literature the
pleasure of passing the evening with this
most agreeable woman. We walked and
conversed upon various literary subjects
till it was dark, when Mrs. Darwin seeming
to be surprised, that the Doctor had not
come home, I offered to take my leave:
but she told me, that I had been expected
for some days, and that a bed had been
prepared for me; I heard some orders
given to the housemaid, who had destined
a different room for my reception from

that which her mistress had upon second thoughts appointed. I perceived that the maid examined me attentively, but I could not guess the reason. When supper was nearly finished, a loud rapping at the door announced the Doctor. There was a bustle in the hall, which made Mrs. Darwin get up and go to the door. Upon her exclaiming, that they were bringing in a dead man, I went to the hall: I saw some persons, directed by one whom I guessed to be Doctor Darwin, carrying a man who appeared motionless.

" He is not dead," said Doctor Darwin. " He is only dead drunk. I found him," continued the Doctor, " nearly suffocated in a ditch; I had him lifted into my carriage, and brought hither, that we might take care of him to night."

Candles came, and what was the surprise of the Doctor, and of Mrs. Darwin, to find that the person whom he had saved was Mrs. Darwin's brother! who, for the first time in his life, as I was assured, had been intoxicated in this manner, and who would undoubtedly have perished, had it not been for Doctor Darwin's humanity.

During this scene I had time to survey my new friend, Doctor Darwin. He was a large man, fat, and rather clumsy; but intelligence and benevolence were painted in his countenance: he had a considerable impediment in his speech, a defect, which is in general painful to others; but the Doctor repaid his auditors so well for making them wait for his wit or his knowledge, that he seldom found them impatient.

When his brother was disposed of, he came to supper, and I thought that he looked at Mrs. Darwin, as if he was somewhat surprised, when he heard that I had passed the whole evening in her company. After she withdrew, he entered into conversation with me upon the carriage that I had made, and upon the remarks that fell from some members of the society to whom I had shewn it. I satisfied his curiosity, and having told him, that my carriage was in the town, and that he could see it whenever he pleased; we talked upon other mechanical subjects, and afterwards on various branches of knowledge, which necessarily produced allusions to classical literature; by these he discovered, that I had received the education of a gentleman.

" Why! I thought," said the Doctor, " that you were only a coachmaker!"— " That was the reason," said I, " that you looked surprised at finding me at supper with Mrs. Darwin. But you see, Doctor, how superior in discernment ladies are even to the most learned gentlemen; I assure you, that I had not been in the room five minutes, before Mrs. Darwin asked me to tea."

The next day I was introduced to some literary persons, who then resided at Lichfield, and among the foremost to Miss Seward. How much of my future life has depended upon this visit to Lichfield! How little could I then foresee, that my having examined and understood the Microcosm at Chester should lead me to a place, and into an acquaintance, which would otherwise, in all human probability, have never fallen within my reach! Miss Seward was at this time in the height of youth and beauty, of an enthusiastic temper, a votary of the muses, and of the most eloquent and brilliant conversation. Our mutual

acquaintance was soon made, and it continued to be for many years of my life a
source of never-failing pleasure. It seems
that Mrs. Darwin had a little pique against
Miss Seward, who had in fact been her
rival with the Doctor. These ladies lived
upon good terms, but there frequently occurred little competitions, which amused
their friends, and enlivened the uniformity,
that so often renders a country town insipid. The evening after my arrival, Mrs.
Darwin invited Miss Seward, and a very
large party of her friends, to supper. I
was placed beside Miss Seward, and a
number of lively sallies escaped her, that
set the table in good humour. I remember, for we frequently remember the merest
trifles which happen at an interesting period of our life, that she repeated some of
Prior's Henry and Emma, of which she
was always fond, and dwelling upon Emma's tenderness, she cited the care that
Emma proposed to take of her lover, if he
were wounded,

" To bind his wounds my finest lawns I'd tear,
  Wash them with tears, and wipe them with my hair."

I acknowledged, that tearing her finest lawns, even in a wild forest, would be a real sacrifice from a fine lady; and that washing wounds with salt water, though a very severe remedy, was thought to be salutary; but I could not think, that wiping them with hair could be either a salutary or an elegant operation. I represented, that the lady, who must have had by her own account a choice of lawns, might have employed some of the coarse sort for this operation, instead of having recourse to her hair. I paid Miss Seward, however, some compliments on her own beautiful tresses, and at that moment the watchful Mrs. Darwin took this opportunity of drinking *Mrs. Edgeworth's health.* Miss Seward's surprise was manifest. But the mirth this unexpected discovery made fell but lightly upon its objects, for Miss Seward, with perfect good humor, turned the laugh in her favor. The next evening the same society reassembled at another house, and for several ensuing evenings I passed my time in different agreeable companies in Lichfield.

Mr. Bolton, of Birmingham, happened

at this time to call upon Doctor Darwin.
I shewed him and a few of his friends
some of those deceptions of Comus, which
I had discovered. They were particularly
à propos, as at that time Mr. Bolton was
making a large number of magnets for
exportation. He asked me to his house on
Snow-Hill, in Birmingham. He was at
this period just going to remove to the
wild heath, which has since been converted
into a garden, interspersed with cheerful
villas, by his talents and energy. Mr.
Bolton very kindly sent a proper person with me through the principal manu-
factories of Birmingham. There, and at
Soho, I became in a few hours intimately
acquainted with many parts of practical
mechanicks, which I could not otherwise
have learned in many months.

At my return to my house at Hare
Hatch, in Berkshire, I applied with fresh
energy to mechanical invention, and to
the execution of various contrivances. I
was now however aware, that invention
must be grievously wasted, where there is
not a competent knowledge of what has
been previously done by others. I there-

fore carefully looked over books of mechanical inventions, wherever I could find
them; and I many times discovered, that
what had appeared to me new, and entirely
my own, had been printed a hundred years
ago.   Even the method of locking carriages in turning, invented by Dr. Darwin
and by me, had been employed in a sailing
carriage described in the " Machines ap-
prouvées" of the French Royal Academy.
I entreat readers who may not have any
taste for subjects of this sort, to pardon me
for thus detailing my mechanical pursuits;
it is necessary to shew the progress of my
mind.

I was riding one day in a country, that
was enclosed by walls of an uncommon
height; and upon its being asserted, that it
would be impossible for a person to leap
such walls, I offered for a wager to produce
a wooden horse, that should carry me safely
over the highest wall in the country.   It
struck me, that, if a machine were made
with eight legs, four only of which should
stand upon the ground at one time; if the
remaining four were raised up into the
body of the machine, and if this body

were divided into two parts, sliding, or rather rolling on cylinders, one of the parts, and the legs belonging to it, might in two efforts be projected over the wall by a person in the machine; and the legs belonging to this part might be let down to the ground, and then the other half of the machine might have its legs drawn up, and be projected over the wall, and so on alternately. This idea by degrees developed itself in my mind, so as to make me perceive, that as one half of the machine was always a road for the other half, and that such a machine never rolled upon the ground, a carriage might be made, which should carry a road for itself. It is already certain, that a carriage moving on an iron rail-way may be drawn with a fourth part of the force requisite to draw it on a common road.

After having made a number of models of my machine, that should carry and lay down its own road, I took out a patent to secure to myself the principle; but the term of my patent has been long since expired, without my having been able to unite to my satisfaction in this machine strength with sufficient lightness, and with regular motion, so as to obtain the advantages I proposed.

As an encouragement to perseverance, I assure my readers, that I never lost sight of this scheme during forty years ; that I have made considerably above one hundred working models upon this principle, in a great variety of forms; and that, although I have not yet been able to accomplish my project, I am still satisfied that it is feasible. The experience, which I have acquired by this industry, has overpaid me for the trifling disappointments I have met with; and I have gained far more in amusement, than I have lost by unsuccessful labor. Indeed the only mortification that affected me was my discovering many years after I had taken out my patent, that the rudiments of my whole scheme were mentioned in an obscure Memoir of the French Academy.

About the year 1769, I made a phaeton of uncommon lightness, and so furnished with springs, that each wheel could rise over any obstacle in its way. This carriage had no perch, and had all the advantages, which are now (in 1808) claimed by the patentee for carriages without perches. In fact, however, a light perch is no disadvantage to a carriage, but on the contrary adds to its security. I sent this carriage to the " Society of Arts &c. ;" and at the same time I sent a model of a very large umbrella for covering haystacks, also of a waggon divided into two parts, each having four wheels, so that the roads never sustained more than half the weight of the common load. By connecting these waggons, and applying the horses to the foremost, the necessity of having an additional driver was prevented, which would be required if horses were put to each part of the divided carriage. The load by these means need not be piled so high as usual.

I sent also to the Society of Arts a machine for cutting turnips, which consisted simply of a circular trough with a chopping knife moving on a pin in the centre, so that the person who worked it had nothing to do but to walk round the circle, and to lift the cutter up and down, as a turner works his paring knife. This was put in competition with the machine for cutting turnips, which is now in common use, and for which the society adjudged to Mr. Edgehill the premium. Very little difference was perceived in the performance of our machines; and I still employ my own, because it can be made any where, of any coarse timber, has but one knife, which can be easily kept from rust, and readily sharpened; in short, it performs nearly as much work as Mr. Edgehill's turnip-cutter, and does not cost one fourth part as much. The machine which I use is a trough on three legs, about five feet long, a foot wide, and a segment of a circle of six feet diameter.

For these models and machines the society presented me with their gold medal in 1769. I laid before that society a machine for measuring the force exerted by horses in drawing ploughs and waggons, and in giving motion to machinery of all sorts. It consisted of a swing-tree bar that contained two levers, to one end of each of which the traces were attached, and the other end compressed a spiral spring. There was an index adapted to the swing-tree bar, on which the longest end of one of the levers pointed out the power, which it exerted to bend the spring. This instrument I explained to the Committee of Mechanics of the Society of Arts, in the year 1771. I went abroad in that year, and during my absence a person belonging to the society obtained credit for exhibiting a similar machine. I took no notice of this transaction, as it was of a trifling nature. I have since frequently tried this instrument, and have found it insufficient. It registers only the greatest exertion of the draft; it does not give the sum of the whole exertion in a given time. There is a method, however, by which the force of draft may be ascertained; I hope to have leisure, to put it into practice, and by these means to determine on the subject of wheel-carriages the discordant opinions of men of science and men of business, or of those who call themselves *practical men*; I also hope to decide the different advantages of various ploughs, and other instruments of husbandry, whether drawn by horses or moved by men. This is a matter of no small importance, considering the enormous expense, to which carriers

and farmers are put, in consequence of erroneous opinions, or of new and unfounded
pretensions to improvement. Were a
course of experiments set on foot on this
subject, upon an extensive and liberal plan,
which should be open to the public at
large, and which should be made upon
full sized carts, ploughs, and waggons, the
common sense and ocular conviction of
all ranks of people would soon put disputes on these points to rest. I mean so
far as relates to such machines as are in use
at present. Whenever any new invention
makes its appearance, it should of course
be submitted to similar trials. Attempts
to impose upon the public would by these
means be rendered abortive.

I should have mentioned in its proper
place, that, in the year 1768, the Society
for the Encouragement of Arts, Manufac-
tures, and Commerce, gave me their silver
medal for a perambulator, which, upon
trial in Coldbath fields, answered with
such precision, as would, from proving too
much, have led to a suspicion of error, and
might have discredited belief in its performance altogether, had not the experi-

ment been witnessed by a large committee of the society. It was entered immediately upon their journals. A mile was first cautiously measured by a common chain—afterwards by the perambulator, and again by the same machine returning. The distance measured, going and returning, did not differ so much as one inch. This instrument is described in the Memoirs of the Society, and in other books. I have found, that upon ground, where it met with no interruption from hedges and ditches, it has never failed with me to measure more accurately, and with much greater expedition, than the chain.

An uncommon instance of plagiarism occurred in Ireland relative to this perambulator. About twenty years after I had received my medal, a person presented my perambulator, without any change in its construction, as his own invention, to the Dublin Society. I saw it there, and observed an inscription with the name—which I forget—of the person who presented it, and these lines from Horace,

" Si quid novisti rectius istis,
    Candidus imperti; si non, his utere mecum."

I have endeavoured to collect together whatever related to mechanics during my residence at Hare Hatch, without attending strictly to the order of time; for it is better to consider the course of one subject uniformly, and then to go back and take up another, than to interrupt the narrative by an ineffectual attempt to preserve strict chronology.

I have not during the account of the foregoing period of my life mentioned my children. I said that I married in 1763. My eldest son was born at Black-Bourton, in Oxfordshire, in 1764. After my return from Ireland in 1765, when I established myself at Hare Hatch, I formed a strong desire to educate my son according to the system of Rousseau. His Emile had made a great impression upon my young mind, as it had done upon the imaginations of many far my superiors in age and understanding. His work had then all the power of novelty, as well as all the charms of eloquence; and when I compared the many plausible ideas it contains, with the obvious deficiencies and absurdities, that I saw in the treatment of children in almost every family, with which I was acquainted, I determined to make a fair trial of Rousseau's system. My wife complied with my wishes, and the body and mind of my son were to be left as much as possible to the education of nature and of accident. I was but twenty-three years old, when I formed this resolution; I steadily pursued it for several years, notwithstanding the opposition with which I was embarrassed by my friends and relations, and the ridicule by which I became immediately assailed on all quarters.

I dressed my son without stockings, with his arms bare, in a jacket and trowsers such as are quite common at present, but which were at that time novel and extraordinary. I succeeded in making him remarkably hardy: I also succeeded in making him fearless of danger, and, what is more difficult, capable of bearing privation of every sort. He had all the virtues of a child bred in the hut of a savage, and all the knowledge of *things*, which could well be acquired at an early age by a boy bred in civilized society. I say knowledge of *things*, for of books he had less knowledge

at four or five years old, than most children have at that age. Of mechanics he
had a clearer conception, and in the application of what he knew more invention,
than any child I had then seen. He was
bold, free, fearless, generous; he had a
ready and keen use of all his senses, and of
his judgment. But he was not disposed
to *obey*: his exertions generally arose from
his own will;.and, though he was what is
commonly called good-tempered and good-
natured, though he generally pleased by
his looks, demeanour, and conversation,
he had too little deference for others, and
he shewed an invincible dislike to control.
With me, he was always what I wished;
with others, he was never any thing but
what he wished to be himself. He was,
by all who saw him, whether of the higher
or lower classes, taken notice of; and by all
considered as very clever. I speak of a
child between seven and eight years old,
and to prevent interruption in my narrative, I here represent the effects of his education from three to eight years old, during
which period I pursued with him Rousseau's
plans.

I now come to what I consider as a new era in my life, the commencement of my acquaintance with Mr. Day. He lived at this time with his father and mother at Barehill, in Berkshire. He came from Oxford during vacation, and hearing that I had been of the same college with him, and a pupil of his tutor Mr. Russell, he came to Hare Hatch to pay me a visit. Mr. Day's exterior was not at that time prepossessing, he seldom combed his raven locks, though he was remarkably fond of washing in the stream. We conversed together for several hours on his first visit, and thus began an acquaintance which was I believe of service to us both. To the day of his death, we continued to live in the most intimate and unvarying friendship—a friendship founded upon mutual esteem, between persons of tastes, habits, pursuits, manners, and connexions totally different. A love of knowledge, and a freedom from that admiration of splendour, which dazzles and enslaves mankind, were the only essential points in which we entirely agreed. Mr. Day was grave and of a melancholy tempe-

rament; I gay and full of " constitutional joy." Mr. Day was not a man of
strong passions,—I was —Mr. Day was
suspicious of the female sex, and averse
to risking his happiness for their charms
or their society.—To a contrary extreme
I was fond of all the happiness, which
they can bestow. He delighted, even in
the company of women, to descant on the
evils brought upon mankind by love: he
used, after enumerating a long and dismal
catalogue, to exclaim with the satiric poet,

" These, and a thousand more, we find :
Ah! fear the thousand yet unnam'd behind."

I used to reply with the Anacreontic
song,

" How I baffle human woes,
Woman, lovely woman knows."

Mr. Day could not refrain from frequently tempting his fate!—and, what
was still more extraordinary, he expected
that, with a person neither formed by nature, nor cultivated by art, to please, he
should win some female wiser than the
rest of her sex, who should feel for him the
most romantic and everlasting attachment

—a paragon, who should forget the follies
and vanities of her sex for him; who

    " Should go clad like our maidens in grey,
       And live in a cottage on love."

These hopes and feelings sprang from
noble and generous motives. Though
armed in adamant against the darts of
beauty, and totally insensible to the power
of accomplishments, he felt, that for an object, which should resemble the image in his
fancy, he could give up fortune, fame, life,
every thing but virtue. It is but justice,
and not the partiality of friendship, that
induces me to assert, that Mr. Day was the
most virtuous human being whom I have
ever known. During three and twenty
years, that we lived in the most perfect
intimacy, I never knew him swerve from
the strictest morality in words or actions.
How far beyond the rigid line of duty
his humanity, universal benevolence, and
unbounded generosity carried him in his
intercourse with mankind, even the un-
reserved friendship, in which he lived with
me or with any other of his friends, could
never enable us to estimate. In the course

of this narrative many instances of his liberality, and of that highmindedness, which distinguished him from other men, will appear. The true patriotism of Mr. Day's mind, unbiassed by the love of popularity; or attachment to party, cannot be fully appreciated even by his writings, though every line he wrote breathes the purest love of his country.

After our first meeting, scarcely a day passed whilst I lived at Hare Hatch without our spending several hours together. On literature of all sorts we conversed, but metaphysics in particular became the subject of our consideration. We differed frequently for months, nay even for years, upon various points; but in time we generally came to the same conclusion. I never was acquainted with any man, who in conversation reasoned so profoundly and so logically, or who stated his arguments with so much eloquence, as Mr. Day.

It was a singular trait of character in my wife, who had never shewn any uneasiness at my intimacy with Sir Francis Delaval, that she should take a strong dislike to Mr. Day. A more dangerous and seductive

companion than the one, or a more moral
and improving companion than the other,
could not be found in England. This jealousy of my friend was a subject of great uneasiness to me. My wife was prudent, domestic, and affectionate; but she was not
of a cheerful temper. She lamented about
trifles; and the lamenting of a female, with
whom we live, does not render home delightful. Still I lived more at home than
is usual with most men of my age. I did
not belong to any club in the neighbourhood; nor did I frequent any assembly,
or the yearly races of Reading or Maidenhead, which were within seven or eight
miles of me. Except paying one visit to Sir
Francis Delaval, three or four short visits
to Birmingham and Lichfield, a visit to my
father in Ireland, and the days necessary
for keeping terms at the Temple, I never
dined or slept from home ten times during
five or six years.

Besides my friendship with Mr. Day, I
about this time formed an intimacy with
Mr. Keir, of Birmingham, a gentleman
well known in the literary world. He had
served abroad, and had obtained the rank

of captain; but, wisely despising the idleness of a soldier's life in time of peace, he
sold out of the army at the peace of Fon-
tainbleau, and turned the energy of his
powerful mind to science, with a view to
make some discovery, by which he might
increase his fortune, and in the pursuit of
which he might find interesting occupation. I became acquainted with Mr. Keir,
at the time when he was employed in trans-
lating Macquer's Dictionary of Chemistry,
a work which was rendered doubly valuable by the notes of the translator. Mr.
Keir accepted an invitation to my house,
where he had leisure to pursue his studies
during several months. By my means he
became acquainted with Mr. Day; and by
means of Mr. Keir I became acquainted
with Dr. Small, of Birmingham, a man
esteemed by all who knew him, and by
all who were admitted to his friendship
beloved with no common enthusiasm. Dr.
Small formed a link, which combined Mr.
Bolton, Mr. Watt, Dr. Darwin, Mr. Wedgwood, Mr. Day, and myself, together—
men of very different characters, but all
devoted to literature and science. This

mutual intimacy has never been broken but by death ; nor have any of the number failed to distinguish themselves in science or literature. Some may think, that I ought with due modesty to except myself.

It is not my object to write the lives of the gentlemen, whom I have named as my particular friends ; but I cannot refrain from noticing the great variety of intellect, which they possessed. Mr. Keir, with his knowledge of the world, and good sense : Dr. Small, with his benevolence and profound sagacity : Wedgwood, with his unceasing industry, experimental variety, and calm investigation : Bolton, with his mobility, quick perception, and bold adventure : Watt, with his strong inventive faculty, undeviating steadiness, and unbounded resource : Darwin, with his imagination, science, and poetical excellence : and Day, with his unwearied research after truth, his integrity and eloquence :—formed altogether such a society, as few men have had the good fortune to live with ; such an assemblage of friends, as fewer still have had the happiness to possess, and keep through life.

At a later period I became acquainted with two friends of Mr. Day's, Mr. Bicknel, and Mr. William Seward, of London. Mr. Seward was prevented by hypochondriacism from much exertion: he however compiled " Biographical Anecdotes" with some reputation. Mr. Bicknel was a man of uncommon abilities. In conjunction with Mr. Day he wrote that beautiful and popular poem, called " *The Dying Negro*," which was published under the name of Mr. Day; nor did I know, till after Mr. Day's death, the precise share, which Mr. Bicknel had in this publication. It is always agreeable to learn what parts of any joint literary production belong respectively to each of its authors. I may therefore mention, that in the " Biographical Dictionary" of Dr. Kippis may be found an analysis of this poem, in which those lines which were written by Mr. Bicknel are pointed out, as they were distinguished in a copy that was in the possession of Mrs. Day.

Mr. Bicknel, beside being a good poet, and a lawyer of the most promising abilities, was a man of nice discrimination, sound judgment, and various conversation. Besides these, whom I may call my intimate friends, I was introduced by Mr. Keir into a society of literary and scientific men, who used formerly to meet once a week at *Jack's* Coffee House, in London, and afterwards at *Young Slaughter's* Coffee House. Without any formal name, this meeting continued for years to be frequented by men of real science, and of distinguished merit. John Hunter was our chairman. Sir Joseph Banks, Solander, Sir C. Blagden, Dr. George Fordyce, Milne, Maskelyne, Captain Cook, Sir G. Shuckburgh, Lord Mulgrave, Smeaton, and Ramsden, were among our numbers. Many other gentlemen of talents belonged to this club, but I mention those only, with whom I was individually acquainted. A society of literary men, and a literary society, may be very different. In the one, men give the result of their serious researches, and detail their deliberate thoughts. In the other, the first hints of discoveries, the current observations, and the mutual collision of ideas, are of important utility. The knowledge of each member of such a society becomes in time disseminated among the whole body, and a certain *esprit de corps,* uncontaminated with jealousy, in some degree combines the talents of numbers to forward the views of a single person. I have felt, ever since I belonged to this society, the advantage of its conversation. I am not a freemason; I cannot, therefore, speak of the initiatory trials, to which a brother is subjected; but in the society of Slaughter's Coffee House we practised every means in our power, except personal insult, to try the temper and understanding of each candidate for admission. Every prejudice, which his profession or situation in life might have led him to cherish, was attacked, exposed to argument and ridicule. The argument was always ingenious, and the ridicule sometimes coarse. This ordeal prevented for some time the aspiration of too numerous candidates; but private attachments at length softened the rigour of probation, the society became too numerous, and too *noble,* and was insensibly dissolved.

A singular instance of the advantage, which arises from a free communication of ideas, happened to a medical gentleman,

one of our members, who had been called
upon to give evidence in a cause of great
moment, and the evidence he gave was
contrary to the opinion of most gentlemen.
of his own profession. On the trial of
Donellan, who was convicted of having
poisoned Sir Theodosius Boughton with
laurel water, the gentleman, to whom I
allude, gave it as his opinion, that such a
quantity of this water, as would kill a dog
instantly, might be swallowed by a man
without any material inconvenience. This
assertion was much questioned both at the
trial and afterwards. At a meeting of our
society, a gentleman attacked it with great
vehemence, and he went so far as to say,
that the opinion was fabricated for the occasion of the trial of Donellan.

A very unbecoming scene began, which
was fortunately interrupted by my calling
upon two members of the club to witness,
that we had heard the medical gentleman
alluded to relate ten years before the
trial, that he had swallowed as much laurel-
water as would have killed twenty dogs.
I added, that I had seen a dog killed at
Dr. Smith's anatomical lectures, at Ox-

ford, by the injection of a single drop of laurel-water. The assailant was quickly put down by my evidence. The simple fact, that the experiment had been made by my medical friend, and communicated to a literary society, many years previous to Donellan's trial, proving, that it could not have been fabricated *for the occasion.*

I have mentioned that Ramsden, the celebrated optician, was of our society. Besides his great mechanical genius, he had a species of invention not quite so creditable, the invention of excuses. He never kept an engagement of any sort, never finished any work punctually, or ever failed to promise what he always failed to perform.

The king (George III.) had bespoke an instrument, which he was peculiarly desirous to obtain; he had allowed Ramsden to name his own time, but, as usual, the work was scarcely begun at the period appointed for delivery; however, when at last it was finished, he took it down to Kew in a postchaise, in a prodigious hurry; and, driving up to the palace gate, he asked if *His Majesty was at home.* The

pages and attendants in waiting expressed
their surprise at such a visit : he however
pertinaciously insisted upon being admitted,
assuring the page, that, if he told the king
that Ramsden was at the gate, His Majesty
would soon shew that he would be glad to
see him. He was right, he was let in, and
was graciously received. His Majesty, after
examining the instrument carefully, of
which he was really a judge, expressed his
satisfaction, and turning gravely to Ramsden, paid him some compliment upon his
punctuality.

"I have been told, Mr. Ramsden," said
the King, "that you are considered to be
the least punctual of any man in England ;
you have brought home this instrument
on the very *day* that was appointed. You
have only mistaken the *year !*"

CHAPTER VIII.

In the spring of 1768, my friend Mr. Day accompanied me to Ireland, on a visit to my father. I took my son with me, leaving my wife and an infant daughter in England. We travelled in my phaeton, and with my own horses, but without a servant. To amuse ourselves on our journey, we agreed that Mr. Day should pass for a very *odd* gentleman, who was travelling about the world to overcome his sorrow for the loss of his wife; he was to be doatingly fond of his son, who was to be a most extraordinary child. We settled that I should pass for his servant and factotum; that whilst I behaved with the utmost civility and attention towards my master, I should behind his back represent him as a humorist and a misanthropist; and that

while he appeared civil, and easily pleased
with common fare and ordinary attendance,
I should give myself all possible airs. We
put our plan in execution at Eccleshall, in
Staffordshire. I drove my indolent master
with great eclat up to the inn, which stands,
or stood, in a broad street. My carriage
had a contrivance for letting off the horses
instantaneously. As soon as I came to the
door, I disengaged the horses in a moment,
before the hostler, for whom I vociferated
with uncommon energy, could obey my
summons. A number of people collected
about us, who gazed at Mr. Day and the
child, sitting composedly in the open carriage without horses. I got out, and with
my hat off, lifted out little master, and
held my arm for my great master, who
descended with much deliberation. When
they had been shewn into a parlour, I
went into the kitchen to order dinner. The
landlady was going up to ask what the
gentleman would please to have; but I
stopped her, and begged that she would
save herself the trouble, as I always ordered every thing. I then bespoke some
cold meat for my master, and a tart for

the child, and with an air of authority desired to see the larder, that I might order
dinner for *myself*. Mrs. Landlady, as she
did not live on one of the great roads near
London, seemed somewhat surprised at
my assurance. Having made my way to
the larder, I ordered every thing that appeared delicate or costly in the house. I
now went into the street, examining the
carriage, sending the hostler one way and
Boots another.

My son, who, though very young, entered into the joke, was looking out of the
dining-room window; and, desirous of
playing some part in the drama, he came
down to me, climbed up into the phaeton,
and then jumped out into my arms from a
considerable distance. He performed various other feats of activity, and these, with
his dress, attracted the notice of those inhabitants of the town who had collected near
the inn; they began to talk to the boy and
to me with eager curiosity. I diverted
them exceedingly with an account of Mr.
Day's misanthropy, and with his various
adventures by sea and land; and thus I
contrived to fix their attention till dinner-

time approached. My master then called for the child, and I carried him in with the utmost haste, and with the most obsequious manner. My master's plain fare was ready, and my delicate dinner was preparing with the utmost nicety in the kitchen, where I condescended to direct the cook, when on a sudden I was accosted by my name in a well known voice. I turned and beheld Doctor Darwin, whom I little expected to meet at this distance from his home, and travelling a road unusual to him. The landlady cast a reproachful look upon me, but, notwithstanding her evident surprise, wanted to persuade us, that she had not been *taken in from the beginning*. " Ah! Doctor," said she, " this gentleman wanted to pass himself upon me for a servant; but I suspected him, notwithstanding all his pretensions. He has ordered every thing good in the house, and I hope he will share his dinner with his friends, who have acknowledged him in his low estate."

Mr. Whitehurst * was with Doctor Dar-

* Author of a " *Theory of the Earth*," &c.

win, and we joined company. Thus began my acquaintance with one of the most simple, unassuming philosophers I have ever known, and thus I had an opportunity of introducing Mr. Day to Doctor Darwin. Their acquaintance did not however this day commence under the most favorable auspices; for as Mr. Day had no taste for mechanicks, he did not join in our conversation for several hours, during which time Doctor Darwin of course took him to be just such a person in reality, as I had represented him to our landlady. Some topic however arose before we parted, on which Mr. Day displayed so much knowledge, feeling, and eloquence, as to captivate the Doctor entirely. He invited Mr. Day to Lichfield, an invitation which led not only to intimacy, but to a very sincere friendship.

After our departure from Eccleshall, Mr. Day and I proceeded without any adventures on our journey to Ireland, and arrived in Dublin. This city was at that time so very different from what it now is, and from what London then was, that it struck Mr. Day with surprise and disgust. The streets were wretchedly paved, and more dirty than can be easily imagined. The poor were squalid, and their tones strangely discordant to an English ear. The hackney coaches, their horses, and still more their drivers, were so far below what were to be seen in London, and were altogether so uncouth, as to increase at every fresh view the astonishment of my friend. As we passed through the country, the hovels in which the poor were lodged, which were then far more wretched than they are at present, or than they have been for the last twenty years, the black tracts of bog, and the unusual smell of the turf fuel, were to him never-ceasing topics of reproach or lamentation. Mr. Day's deep-seated prejudice in favor of savage life was somewhat shaken by this view of want and misery, which philosophers of a certain class in London and Paris chose at that time to dignify by the name of simplicity. The modes of living in the houses of the gentry were much the same in Ireland as in England. This surprised my friend: he observed, that, if there was any difference,

it was that people of similar fortune did
not restrain themselves equally in both
countries to the same prudent economy;
but that every gentleman in Ireland, of two
or three thousand pounds a year, lived in
a certain degree of luxury and show, that
would be thought presumptuous in persons
of the same fortune in England.

On our journey to my father's house, I
had occasion to vote at a contested election
in one of the counties through which we
passed. Here a scene of noise, riot, confusion, and drunkenness was exhibited, not
superior indeed as to depravity and folly,
but of a character or manner so different
from what my friend had ever seen in his
own country, that he fell into a profound
melancholy. No beds were to be had at
the inn, or stabling for my horses; but if
we and they could have eaten beef and
drunk claret in the streets, we might have
had as much, or ten times as much, as
we could have swallowed. The gentleman,
for whom I intended to vote, at length
billeted me at a clergyman's house near
the town, where we were most comfortably
taken care of. The next morning I went

early to the hustings, that I might give my vote, and proceed without delay upon my journey; I was the first person who offered himself to be polled. None of the candidates knew me; and the sheriff, to whom also I was a stranger, inquired from the lawyers and from those near him, who I was; but could not discover. A law had been just made to expedite elections by prohibiting the examination of voters as to their freeholds; and the sheriff was obliged at his peril to receive every vote, that was sanctioned by the oath of the voter. This law had not yet been enforced at any election, and I was determined to exercise my steadiness, by adhering strictly to the letter of the Act of Parliament, which I had in my pocket. I accordingly presented myself to the sheriff with a tender of my vote. He asked me various questions, which I thought irrelevant; but which the lawyers on all sides, for there were four candidates, insisted I should answer.—I refused—and, producing the act, asserted my right of voting, upon making the affidavit therein specified. The sheriff began to hesitate; the lawyers began to fear, that a rule might be made relative to the future proceedings, which might injure their importance; and in consequence I was actually admitted to vote without giving any explanations. By this time my friend, the father of one of the candidates, came upon the hustings, and told the sheriff, that he knew I had an estate worth some hundreds a year in his county. The lawyers laughed, and I laughed along with them, and retired well satisfied with this essay of my firmness among numbers.

In these countries it is essential to a man of abilities, to be able to think and speak in public assemblies. Yet we often see, that men who can fight, and who in private are rather impudent than modest, frequently find, that their presence of mind forsakes them when they are to speak or decide before numbers. Habit sometimes renders this disposition so unconquerable, that men of experience and of great talents feel themselves panic struck in public assemblies, if they attempt to deliver their sentiments.

Mr. Day and I proceeded to Edgeworth-Town, where my father and sister resided.

To my father, Mr. Day appeared a very
singular personage. My father had not
been used to think, that a man without
certain conventional manners could be entitled to much consideration; and though
he could not avoid being struck by Mr.
Day's conversation, and by his strong
powers of reasoning, yet my father conceived a violent prejudice against him, in
consequence of something in his manner of
eating and sitting at table, which appeared
unsuitable to his rank in life. Mr. Day
on his side smiled with philosophic indifference at these prejudices in favor of politeness, and seemed to undervalue the
understanding of him, who set such high
importance upon external appearance. My
sister, who was not less imbued than my
father with a prepossession in favor of
good breeding, stood aloof; while my friend
preserved an awful distance from a woman,
whom he was inclined to consider as a
confirmed fine lady, a sort of being for
which he had a feeling of something like
horror. My sister's easy manners, and
agreeable conversation, in a few weeks
began nevertheless to unbend the stub-

bornness of Mr. Day's stoicism. She perceived his excellent sense, and humane and generous disposition, and I perceived that they were disposed to approach. The lady however could not be reasoned out of her aristocratic habits, which she defended with so much wit and vivacity, that in company she had always the advantage; but when I was the only auditor, Mr. Day's eloquence prevailed. My sister had a taste for the beauties of nature, for literature, and for good conversation; she was far from being deficient in information on those subjects which had engaged his attention, and there were some points on which they had similar sentiments.

Before three months were at an end, Mr. Day became her avowed admirer, and my sister was prevailed upon to acknowledge, that, if the gentleman continued for a year in the same mind, and could in that time make his appearance becoming a man of his situation in life, she might be induced to give him her hand. I was commissioned to speak to my father. He could make no objection to Mr. Day's morals or fortune; but he could not repress his astonishment at his daughter's descent from the aristocratic heights, to which she had been accustomed.

My friend and I quitted Ireland in autumn, and returned to Hare Hatch. I left my sister studying metaphysics, which Mr. Day had recommended to her; and he, in hopes of pleasing her, went to London to study the graces. I believe that about this time Mr. Day entered the Temple; but I do not distinctly recollect the order of the unimportant events, which passed during the year 1769. For my own part I made several excursions to Lichfield and Birmingham, to visit my friends Dr. Small, Mr. Keir, and Dr. Darwin. I eagerly cultivated their society, and I amused myself with mechanics, but I fear that I did not add much to my stock of literature.

In one of my journeys from Hare Hatch to Birmingham, I accidentally met with a person, whom I as a mechanick, had a curiosity to see. This was a sailor, who had amused London with a singular exhibition of dexterity. He was called *Jack the Darter*. He threw his darts, which consisted of thin rods of deal, of about half an inch in diameter, and of

a yard long, to an amazing height and distance; for instance, he threw them over
what was then called the New Church in
the Strand. Of this feat I had heard, but I
entertained some doubts upon the subject;
I had inquired from my friends where this
man could be found, but had not been able
to discover him. As I was driving towards
Birmingham in an open carriage of a singular construction, I overtook a man, who
walked remarkably fast, but who stopped
as I passed him, and eyed my equipage
with uncommon curiosity. There was
something in his manner, that made me
speak to him; and, from the sort of questions he asked about my carriage, I found
that he was a clever fellow. I soon learned,
that he had walked over the greatest part
of England, and that he was perfectly acquainted with London. It came into my
head to inquire, whether he had ever seen
the exhibition, about which I was so desirous to be informed.

" Lord! Sir," said he, " I am, myself,
Jack the Darter." He had a roll of brown
paper in his hand, which he unfolded, and
soon produced a bundle of the light deal

sticks, which he had the power of darting
to such a distance. He readily consented
to gratify my curiosity, and after he had
thrown some of them to a prodigious height,
I asked him to throw some of them hori-
zontally. At the first trial he threw one
of them eighty yards with great ease. I
observed, that he coiled a small string
round the stick, by which he gave it a ro-
tary motion, that preserved it from alter-
ing its course; and at the same time it allowed the arm, which threw it, time to
exercise its whole force.

If any thing be simply thrown from the
hand, it is clear, that it can acquire no
greater velocity than that of the hand which
throws it; but if the body, that is thrown,
passes through a greater space than the
hand, whilst the hand continues to communicate motion to the body to be impelled,
the body will acquire a velocity nearly
double to that of the hand which throws it.
The ancients were aware of this, and they
wrapped a thong of leather round their
javelins, by which they could throw them
with additional violence. This invention
did not, I believe, belong to the Greeks; nor

do I remember its being mentioned by Homer or Xenophon. It was in use among the Romans; but at what time it was introduced or laid aside I know not. Whoever is acquainted with the science of projectiles will perceive, that this invention is well worthy of their attention.

After having satisfied my curiosity about Jack the Darter, I proceeded to Birmingham. I mentioned, that I travelled in a carriage of a singular construction. It was a one-wheeled chaise, which I had had made for the purpose of going conveniently in narrow roads. It was made fast by shafts to the horse's sides, and was furnished with two weights or counterpoises, that hung below the shafts. The seat was not more than eight and twenty or thirty inches from the ground, in order to bring the centre of gravity of the whole as low as possible. The foot-board turned upon hinges, fastened to the shafts, so that when it met with any obstacle it gave way, and my legs were warned to lift themselves up. In going through water my legs were secured by leathers, which folded up like the sides of bellows; by this means I was pretty

safe from wet.   On my road to Birmingham I passed through Long-Compton, in
Warwickshire, on a Sunday   The people
were returning from church, and numbers
stopped to gaze at me.   There is or was
a shallow ford near the town, over which
there was a very narrow bridge for horse
and foot passengers, but not sufficiently
wide for waggons or chaises.   Towards
this bridge I drove.   The people, not perceiving the structure of my one-wheeled
vehicle, called to me with great eagerness
to warn me, that the bridge was too narrow for carriages.   I had an excellent
horse, which went so fast as to give but
little time for examination.   The louder
they called, the faster I drove, and when I
had passed the bridge, they shouted after
me with surprise.   I got on to Shipston
upon Stour ; but, before I had dined there,
I found that my fame had overtaken me.
My carriage was put into a coach-house,
so that those who came from Long-Compton, not seeing it, did not recognise me ; I
therefore had an opportunity of hearing all
the exaggerations and strange conjectures,
which were made by those who related my

passage over the narrow bridge. There
were posts on the bridge, to prevent, as I
suppose, more than one horseman from
passing at once. Some of the spectators
asserted, that my carriage had gone over
these posts; others said that it had not
wheels, which was indeed literally true;
but they meant to say that it was without
any wheel. Some were sure that no carriage ever went so fast; and all agreed,
that at the end of the bridge, where the
floods had laid the road for some way
under water, my carriage swam on the
surface of the water

This year when I went to London to
keep my last terms, 1 made a first attempt
to speak in a public assembly. My friend
Mr. Day had about this time made a
speech to the Westminster electors, which
had been much applauded. He and our
common friend Mr. Bicknel were in town
living together; as we were going home
one night through the streets of London,
Mr. Bicknel expressed much surprise at
Mr. Day's being able to face such a multi-
tude, as that which he had encountered on
the hustings; and was astonished, that he

could speak with tolerable facility even an harangue which had been prepared at leisure. I affirmed, that a man, who had always been used to express himself in good language, could not be much at a loss in speaking upon any common subject in public; I said that the magick of numbers, like all other magick, ceased to have power the moment when its influence was discredited. "But do you think," said Mr. Bicknel, "that you could get up, without previous preparation, on any given common subject, and speak before a great number of people?" I answered in the affirmative: and I said, that, supposing it was a subject with which I was tolerably well acquainted, I would undertake to make the audience laugh at me, hiss me, and applaud me.

"You may try now," said Mr. Bicknel, "for we are within fifty yards of Coach-makers' hall, where a debating society is at this moment in full assembly." I determined to make the trial. We found the room extremely crowded, not indeed with a very brilliant assembly; but still a crowd is a crowd. The subject under consideration was the influence of female manners upon society. The moment that the speaker on his legs had finished, I rose, and, catching the president's eye, I began, as soon as silence permitted me, to address him thus, stammering at every syllable, " Mr. Pr—— Pr—— Pr—— Mr. Pre —— Pre —— Pre —— Pres—— President,"—stuttering with prodigious violence. Voices on every side expressed wonder at the folly of a man's attempting to speak in public, who must be, or ought to be, conscious, that he could not speak articulately in private. After a few moments of compassion, the ridicule of the spectacle which I afforded struck the assembly with one common consent, and a universal laugh shook the hall. When it began to subside, I appeared to recover myself by degrees, and then, in fluent language, I descanted upon the follies of fashion, and upon the sacrifices made to its despotick power by the fair sex. I observed, that my audience was principally composed of ordinary mechanicks, shoemakers and staymakers in particular. The gallery was full of females, and before

such an audience my second undertaking
was not very difficult to be accomplished.
I inveighed with great vehemence against
the indecency, with which ladies sometimes
submitted their persons to the familiar
touch of rude mechanicks: I described a
delicate lady holding out her leg and foot
to a shoemaker: I marked the contrast
between his black hands and leathern
smell, and the white silken foot and per-
fumed odour of a modern belle: I described, perhaps too accurately, the operations of an Italian staymaker. This was at
a time when stays of a peculiar kind were
in fashion. My audience began to murmur,
and as I proceeded, their feelings broke
out into the strongest marks of indignation. I was hissed, and nearly put down.
I however kept my countenance, and gradually receding from the tone I had taken,
I pronounced a panegyric upon women,
on their beauty, modesty, prudence, and
unlimited sway over our sex. The ladies
in the gallery got up, leaned over the rail-
ing, and shewed by their attention and
their smiles, that I had obtained their approbation. The president's hammer stood

suspended, whilst I proceeded in this strain, till by abusing all foreigners, and praising every thing that was English, I obtained unbounded applause from the whole audience. I never returned to this nursery of eloquence: but I am satisfied, that this trial of myself gave me confidence, when I had occasion many years afterwards to address a public assembly.

It is useful to men to try their powers, but teaching boys to declaim is in general hurtful; as it gives them a pedantic air, and inclines them to think, that, so long as they can go on with fluency, and in language tolerably appropriate, they are *speaking well.* This treacherous facility prevents them from perceiving, that their knowledge is superficial; and they never suspect the inevitable truth, that they must be neglected, unless they have something useful to communicate.

Before the winter of the year in which they parted had elapsed, Mr. Day and my sister discovered, what all their friends had seen from the beginning of their acquaintance, that they were not suited to each other. Fortunately this discovery was made, before my friend had become so far attached, as to render my sister's final decision a source of deep, or at least of permanent regret. Mr. Day suffered indeed at the time, but his heart and his pride soon recovered, and he determined to put in practice a scheme, which had long occupied his imagination. This was no common project, but a design more romantic than any which we find in novels.

Mr. Day resolved to breed up two girls, as equally as possible, under his own eye; hoping that they might be companions to each other while they were children, and that, before they grew up to be women, he might be able to decide, which of them would be most agreeable to himself for a wife. I was not with my friend, when he and Mr. Bicknel selected from a number of orphans*, one of remarkably promising appearance. It was necessary, that the girl should be bound apprentice to some *married* man. I was the person, whom Mr. Day named, and to me Sabrina Sidney was apprenticed. Mr. Day called her Sabrina

* See Mr. Keir's Life of Day.

from the river Severn, and Sidney from his favorite, Algernon Sidney. On his return to London, he presented to me the little ward, who had been thus bound to me without my knowledge. I had such well merited confidence in Mr. Day, that I felt no repugnance against his being entrusted with the care of a girl, who had been thus put incidentally under my protection. In a few days he went to the Foundling Hospital, in London, and chose another girl, to whom he gave the name of Lucretia. He placed his wards in a widow's house, in some court near Chancery Lane, and immediately applied himself to their education. They were eleven and twelve years old, good humored, and well disposed. Mr. Day's kindness soon made them willing to conduct themselves according to his directions. But a lodging in London was not a convenient or an agreeable scene, for such a plan as he intended to pursue; he therefore determined to take his pupils out of England, that he might avoid the inquiries and curiosity of his acquaintance. He accordingly removed from London, and shortly afterwards he sailed to France. I do not remember many of the circumstances of this journey. I know only, that he resided a considerable time at Avignon, where he excited much surprise by his mode of life, and by his opinions. But his simplicity of conduct, strict morality, uncommon generosity, and excellent understanding, soon removed these impressions: and both he and his pupils were treated with kindness and civility by the principal people in Avignon.

Mr. Day had as large a portion of national prejudice in favor of the people of England, and against the French, as any man of sense could have; it was therefore something strange, that he should take two young girls to that country, one of whom he destined to be his wife; but, upon reflection, it appears, that Mr. Day had a considerable advantage in being in this situation. From their total ignorance of the French language, an ignorance, which he took no pains to remove, his pupils were not exposed to any impertinent interference; and as that knowledge of the world, from which he wished to preserve them, was at one entrance quite shut out, he had

their minds entirely open to such ideas
and sentiments, and such only, as he desired to implant. Mr. Day had an unconquerable horror of the empire of fashion
over the minds of women; simplicity, perfect innocence, and attachment to himself,
were at that time the only qualifications
which he desired in a wife. He was not
perhaps sufficiently aware, that ignorance
is not necessary to preserve innocence: for
this reason he was not anxious to cultivate
the understandings of his pupils. He
taught them by slow degrees to read and
write; by continually talking to them, by
reasoning, which appeared to me above
their comprehension, and by ridicule, the
taste for which might afterwards be turned
against himself, he endeavoured to imbue
them with a deep hatred for dress, and
luxury, and fine people, and fashion, and
titles. At his return to England, which
happened, I believe, when I was out of
that country, he parted with one of his
pupils, finding her invincibly stupid, or at
the best not disposed to follow his regimen.
He gave her three or four hundred pounds,
which soon procured her a husband, who

was a small shopkeeper. In this situation
she went on contentedly, was happy, and
made her husband happy, and is, perhaps,
at this moment, comfortably seated with
some of her grandchildren on her knees.
His other pupil, Sabrina Sidney, was, at
Mr. Day's return from France, a very
pleasing girl of thirteen. Her countenance
was engaging. She had fine auburn hair,
that hung in natural ringlets on her neck;
a beauty, which was then more striking,
because other people wore enormous quantities of powder and pomatum. Her long
eye-lashes, and eyes expressive of sweet-
ness, interested all who saw her, and the
uncommon melody of her voice made a
favorable impression upon every person to
whom she spoke. I was curious to see how
my friend's philosophic romance would end.

Note by the Editor.

Two of Mr. Day's letters from Avignon, written when
he was scarcely twenty, may, perhaps, amuse the reader.
His expressions of contempt and horror of French society
must not be taken literally or seriously. Mr. Day did not
understand French sufficiently at this time, to judge of foreign conversation. His exaggerated opinion of Rousseau

was recanted, as will appear by one of his letters after he attained to years of discretion. The following are given merely as specimens of his early style, and as almost the only instances of gaiety of manner, which ever appeared in his correspondence.

Avignon, November, 1769.

BEHOLD me at Avignon, full six hundred and fifty miles, three quarters, and one furlong, from Barehill, (N. B. by the well known rules of addition and subtraction, you may by this means calculate my distance from London, Reading, &c.) and yet, by heavens! I am alive! and what is more, tolerably well; *vivit?—imo vero etiam in senatum venit!* Were I to relate the stagecoaches I have travelled in, the post-boys I have talked big to, (nay, I have gone so far as to say *sacre Dieu!)* the inns I have lain at, the rivers I have passed, with no more than a three-quarters of an inch plank between me and destruction, I should make you shudder! Happy, *O terque quaterque beate,* are you whom fate permits to lead an easy, safe, and inglorious life, far from the toils, the dangers, of us who travel to see the wonders of the world. Thus much I thought proper to observe *en passant.*

And yet, my friend, Providence, that impartial distributor of good and evil, to every station of life allots its peculiar pleasures, as well as its peculiar pains. Had I staid at home, perhaps at this moment I might be in a warm comfortable room, calculating the vibrations of your wooden horse's legs; but should I, my friend, should I have been what I now am,—the traveller, the polite scholar, and the fine gentleman? Should I have worn a laced coat? Should I have seen a governor, or an intendant, or who knows what is yet reserved for me by destiny?

Seriously—It is seldom that we are enabled with impartiality to estimate either the advantages or the disadvantages of our own situation, relative to that of other nations, or other individuals. Were you to believe some of our travellers, to find a sample of hell you must travel in France—bad roads, carriages, provisions, beds, and every other inconvenience which a man can bear and live: were you to believe others, there is no misery except in England, no happiness but in France. Heaven and hell separated by an extent of sea scarcely seven leagues. The first you always laughed at, and with reason: the second, perhaps, you may still believe. To destroy your prejudices against your own country and its inhabitants, and to convince you, that your own literature is by much the most eligible, cannot, I think, be doing you any considerable harm. Me, perhaps, you may believe, or rather believe my writings, my sentiments, if you find a spirit of moderation and candor, without bitterness, and without violence.

There is indubitably among the French a greater spirit of dissipation than among the English: they are accustomed to no kind of employment, to no kind of attention; their mornings are spent in dress and in sauntering about, and their afternoons in visits. If true happiness consists in perfect vacuity, they certainly have the advantage of us. I have been introduced into all the polite assemblies—I know something of their manner of life, at least the outward and visible signs. In their visiting rooms, you see a number of beings lolling, walking, standing, yawning, talking of the same trifling subjects, which you would hear discussed in England with the same indifference, till the happy moment arrives, which sets them down to the gaming table ; I know not that the French are more indifferent to the loss of their money, or less agitated by the common passions which attend the gaming table, than the English ; if they are, I am sure the countenance is no index to the heart. I should be sorry if our Creator had made our happiness depend upon the gaming table. Indeed the rest of mankind would have some cause to complain as well as myself, if this had been so ordained. If you go into their coffee-houses, you find a number of idle people playing at dice, sitting round a stove doing nothing, gaping, yawning, getting up, and sitting down again. There is another species of pleasure in France, which consists in intrigue * * * * * * * * * * * * * * * * Were I to wander like Cain over the face of the earth, there is no doubt but I should prefer France, since it is so much easier for a stranger to get into society here, than in England ; there is also no manner of doubt, but that there is a more general spirit of politeness among the French than among us, that is, a man runs less hazard of being affronted, or meeting with any kind of incivility or positive rudeness; no wonder, for the consequence is death, or infamy, both to the aggressor and the injured ; but were I to settle, I think no man of common sense would hesitate a moment between the two countries ; in England one enjoys all the comforts, all the real advantages, all the connexions of life, let me add all the conveniences, to a much greater advantage than here.

I am settled *chez M. Frédéric, vis à vis la Madeleine, Avignon.* I hope you will write to me as soon as you can : in your letter let me hear of nothing but your boy, your wooden horse, and other domestic occurrences—to me they will be the most agreeable subjects in nature. Every thing belonging to me goes on well, give my love to Dick, and my best respects to your wife. Have you got a house yet?—have you got a patent?—a title?—a fortune?—a child?—a medal?—a new chaise?

Vale,

THOMAS DAY.

1769.

I HAD thought it impossible, that any thing should so far overcome the aversion you entertain to letters, as to make you write a whole sheet of foolscap, and, what is more, in a legible hand. I began to doubt, whether you had not been guilty of plagiary, and bribed some fair lady; as the booksellers paid Dryden to compose so many sheets a winter. I am, however, much obliged by your having made the exception in my favor. When every boarding-school Miss can consume whole reams without a single sentiment or idea to express, it would be disgraceful if two philosophers, and, what is more, two real friends, should be unable to fill a sheet of paper once in six weeks, more especially when we are separated by such a tract of country.

Οὐριά τι σκιόεντα, θάλασσά τι ἠχήεσσα·

Accept a miracle instead of wit,
A line of Homer in Greek letters writ.

That gaiety, my friend, which you remark in my letter, is neither an effect of French, nor of the recovery of my health: it is an effect of either a constitutional philosophy, or of habit to make a jest, at least to others, of what is most disagreeable to me. For be assured no one circumstance of life was ever half so much so, as my residence in France. Dr. Small was little acquainted either with me or the French, when he advised me to a change of climate for the sake of engaging and amusing my mind; instead of giving myself up more to dissipation, I really assure you, I never read or thought more in my whole life, than I do at present.

Consider from how many subjects of English conversation the French are entirely excluded. Politics, that polite source of dispute with us, is with them unknown; with agriculture, which employs so much of the attention of our country gentlemen, they never meddle : add to this the disadvantage of their climate, which, being almost constantly serene, gives them no employment in remarking its variations. Whether it is the effect of custom, prejudice, or both together, I know not, but I begin to regret the fogs and showers of Old England. Nothing can be more ignorant than those of the French Nobility whom I have seen; so far from finding any taste for any kind of science among them, I really have not heard enough in any company in France to persuade me, that any one person who composed it could even read. Upon the whole I conclude, from all I see, and from all I hear, that the general taste for foreign countries, and contempt for their own, so prevalent amongst the English, can proceed from no other principle, than that perversion which makes women prefer lap-dogs to their children, or which makes our sex prefer the company of profligate to modest women, and every species of vice and folly to the more easy dictates of common sense. It is possible, that, deceived by my own prejudices and attachments to men, to places, to opinions, I may judge erroneously ; it is your business to inform yourself of the truth of facts, and then to deduce your own conclusions. Attached entirely to exteriors, enslaved by their king and by women, manliness of sentiment and strength of reason appear to be entirely unknown in this country; to dress, to dance, to sing, to " tend the Fair," are the occupations of a Frenchman's life; their prejudices, which make them consider their inferiors less as men than beasts of burthen, make them entirely indifferent to their wants, their miseries; a profusion in dress, in equipage, usurps the place of love and of generosity. The women, brought up in convents, or formed under the care of gouvernantes and servants, when they enter into the world, bring prejudice, extravagance, and coquetry to their husbands: no laws, nor the force of the religion they are bigoted to, can restrain them; the feeble ties of modesty, decorum, or shame, are unknown—a universal infidelity prevails; the men can feel nothing but indifference for their nominal wives; hence all the ties of nature are broken through, all the sweet connexions of domestic life unknown—husband, wife, father, son, and brother, are words without meaning, *Vox et preterea nihil.* But the most disgusting sight of all is to see that sex, whose weakness of body, and imbecility of mind, can only entitle them to our compassion and indulgence, assuming an unnatural dominion, and regulating the customs, the manners, the lives and opinions of the other sex, by their own caprices, weakness, and ignorance.

But, after all, I think it will be of some advantage to me to have been in France: I flatter myself, that by going into company here, and a little observation, I have so well matured the instructions I received first from you, that I shall have upon my return (or at least know how to assume) sufficient impertinence, loquacity, vanity, and fine clothes, to set up with some degree of success for the character of a *polite* man. Apropos I have bought a fine gold waistcoat for you, but I am apprehensive of its being seized when I come back, therefore I should be glad if you would take the trouble of inquiring, and write me word in your next letter, what are the laws in that respect.

You inquire after my pupils: I am not disappointed in any one respect. I am more attached to, and more convinced of the truths of my principles than ever. I am very sure the company of these children has preserved me from a great many melancholy hours. I have made them, in respect to temper, two such girls, as, I may perhaps say without vanity, you have never seen at the same age. They have never given me a moment's trouble throughout the voyage, are always contented, and think nothing so agreeable as waiting upon me (no moderate convenience for a lazy man): perhaps it may divert you to see an original letter from Miss Sabrina Sydney, word for word dictated by herself:—" Dear Mr. Edgeworth, I am glad to hear you are well, and your little boy—I love Mr. Day dearly, and Lucretia—I am learning to write—I do not like France so well as England—the people are very brown, they dress very oddly—the climate is very good here. I hope I shall have more sense *against* I come to England—I know how to make a circle and an *equilateral* triangle—I know the cause of night and day, winter and summer. I love Mr. Day best in the world, Mr. Bicknell next, and you next." —All this is, I believe, a faithful display of her heart and head.

I can't say I find any benefit from the change of climate, I think I am just the same, *Levius fit patientia quicquid corrigere est nefas.* I know of no consolations we can derive from philosophy against the absolute power of disease. This simple sentiment supports me, That we must all yield to necessity, and, while we live, the less we employ our minds upon our own infirmities, the easier we shall be. When we have lost both the hope of cure and happiness, *Die soon, oh, Fairy's son!* I have allotted myself a kind of task in life, till the performance of which, without the last necessity, I will not retire to rest. After all, there is so little certainty in human affairs, so small an interval between good and evil, happiness and misery, so much, and such unexpected misery amongst mankind, that no one, who considers, will suffer himself to be strongly attached to life. You and I agree perfectly in our fundamental principles. Health of body and employment—moderation of the passions—agreeable society and good temper, *hoc satis est donasse Jovem.* Were all the books in the world to be destroyed, except scientific books (which I except, not to affront you) the second book I should wish to save, after the Bible, would be Rousseau's Emilius. It is indeed a most extraordinary work—the more I read, the more I admire—Rousseau alone, with a perspicuity more than mortal, has been able at once to look through the human heart, and discover the secret sources and combinations of the passions. Every page is big with important truth. In respect to your child, I know of only one danger, which is, that you may enlarge his ideas too fast. To yield without murmuring to necessity, to exert properly the faculties of nature, to be unbiassed by prejudice, are the simple foundations of every thing that is great, good, sublime—" Excellent Rousseau!" first of humankind! Behold a system, which, preserving to man all the faculties, and the excellences, and the liberty of his nature, preserves a *medium* between the brutality and ignorance of a savage, and the corruptions of society! Remember, it will never be too late to enlighten the understanding; but that a single error, like a drop of poison, contaminates the whole. Never trouble yourself about Dick's reading and writing, he will learn it, sooner or later, if you let him alone; and there is no danger, except that the people of Henley may call him a dunce.

Do not be surprised at the enormous length of this letter, or think I expect as much in return: when we exchange ore for metal, he is not always a loser, who receives the least in weight.

Faites mes complimens à Madame votre femme—Mademoiselle votre sœur, &c.

Oh, my dear friend, you'd be quite surprised to see me now: Oh Lord! I am quite another thing to what I was—I *talks* French like any thing; I wears a velvet coat, and a fine waistcoat, all over gold, and dresses quite *comme il faut :* and trips about with my hat under my arm, and " *Serviteur Monsieur !*" and " *J'ai l'honneur Madame,*" &c. Oh dear, it's charming upon my soul !—good night— my paper's out, and I must *dress for the concert.* I pity you poor country puts, that see nothing of the world, and, when I return, will try to teach you how to behave.

THOMAS DAY.

CHAPTER IX.

———◆———

[1769-70.

BEFORE Mr. Day returned from France, and whilst I was still living at Hare Hatch, I received an account, that my father was dangerously ill; I immediately set off for Ireland, and found him in Dublin, suffering under the hands of surgeons, from the effects of a slight cut in one of his toes; he languished for three or four months, bore his illness with resignation to the will of Providence, and died with firm hope of happiness in Heaven. My father, (Richard Edgeworth), was in his seventieth year when he died. He had always set an example to his children of perfect morality, and of unaffected piety. Upright, honourable, sincere, and sweet tempered, he was loved and respected by people of all ranks, with whom he was connected. He was in the Irish parliament for five and twenty years, and

I have reason to believe, that he never gave a vote contrary to his conviction; an instance of virtue, in those days, not very common. That he was in a few instances persuaded by his friends, that what was for their interest was for the interest of the state, might, I believe, be the case; but no corrupt motive could be even suspected, for he never had, during his whole life, any pension or emolument; nor do I know of the smallest favour, that he ever received for any person from government, though he generally voted with what was called the court. He was twice offered, but he declined accepting, a baronetage, to which he had a claim as ancient as the time of James the First, when a patent was prepared for his great grandfather, Francis Edgeworth, who was then clerk of the Hanaper.

My father's understanding was sound and clear, and he was reputed to be a good chamber-council; but his health, or his inclination, did not accord with the laborious practice of his profession; he had retired, soon after his marriage, to the country, and applied himself, by the care and management of his estate, to pro-

vide for his family. Early difficulties had
accustomed him to laborious business at
the desk, and to uncommon accuracy.
These habits continued to influence him
during his whole life, so much, that in the
course of his business he used to give and
take a receipt for sums even under a pound,
at full legal length ; and, to avoid future disputes, he wrote the day of the month and
the year of our Lord in letters. This was
pure habit, and not weakness of mind ; it
is mentioned here, among many instances I
have adduced, to point out the force of
early education.

After my father's death, finding myself
in possession of an estate, which rendered it
no longer necessary for me to pursue my
profession as a barrister, I was not called to
the bar, though I had completed all my
terms. I was not anxious to obtain wealth,
nor was I ambitious, therefore I had no motive to excite me to pursue a laborious profession, that did not suit my taste, which
was for domestic life, and scientific pursuits.

It had been my father's hope, in educating me for the law, that I should distinguish myself in a profession, for which he

had a singular veneration. Had he perfectly understood my disposition, he might
easily have led me to pursue the course,
which he ardently desired. But notwithstanding his having the kindest intentions,
he made some mistakes in the management
of my mind. Seeing that I was of a lively
disposition, he concluded, that I must be
averse to the study of the law; and he fan-
cied, that my love for what is commonly
called *pleasure* would necessarily prevent
me from all serious application. In guard-
ing against this, and in haste to make a
lawyer of me, he at my first going to Oxford, in spite of the honourable warning he
had received, put me under the care of his
friend Mr. Elers. In his house, it is true,
I was free from the common excitements
to dissipation; but I was there, in my first
vacations, secluded from the world, and exposed to a temptation, which to my inexperienced youth, and naturally strong passions, proved irresistible. Had not my father been prepossessed by the idea of my
volatility, and had he placed me in situations where my ambition might have been
early excited, he would probably have suc-

ceeded in his views. It was at that time easy in Ireland, for a young barrister of any talents, and whose family had some influence in their own county, to make way into the Irish parliament. Had parliamentary eminence been held up to me, I should probably have studied hard, to enable myself to follow public business. My father might have encouraged me by pointing out, that I was not entirely deficient in the qualities necessary for a public speaker: I had been always accustomed in my own family to hear good language; I had been in the habit of paying attention to my own manner of speaking; I had read a number of good English authors, and had an ardent taste for the acquisition of knowledge, with a desire to excel, which might have been directed into any path of what is commonly called ambition.

I here simply point out what means would, I believe, have succeeded in securing the objects for me, which my father wished; whether, if I had followed public business, I should have been so happy, in such easy circumstances, or possessed of the same independence of character, as I am at present, it is impossible to determine. But of this I am sure, and I say it with gratitude to Providence, that I have during a long life been in easy circumstances, independent, and happy:— that I have had leisure to judge of right and wrong, to cultivate whatever talents I received from God, and to educate a great number of children, some of whom have already reflected honour upon their father.

The only advantage, which the world can gain from the publication of the lives of individuals, is the knowledge of the circumstances that tend to the formation of character, or of those which influence the happiness of life: on these, therefore, I dwell.

In consequence of my early marriage, I had for many years of my youth been confined, during my father's life, to a small income. Even at that time I was fully sensible, that this was of advantage to my character. It induced the habit of submitting to privations, and of turning aside from allurements to expense. My wife was a good economist. I never was fond of money, nor was I extravagant. I had no expensive tastes, and now, at sixty-five, I feel as

little disposed to love money, as I was at
six and twenty : nor am I more inclined to
wish for superfluities ;

    " Houses, horses, pictures, planting,
    " *No* cruel something *e'er* was wasting."

Hence I was enabled early to live upon
a little, and, as I grew older, with a moderate fortune I made myself affluent by confining my wishes within bounds something
less extended than my income.

Since the time of which I write, the people of Ireland have improved more than
any other people in Europe.　At that period, there existed among a class just above
those who worked for bread a strong disposition to cunning, and an eager desire to
take advantage of the easiness and good
nature of their superiors.　I experienced
this soon after I came into possession of my
father's estate.　I felt much disposed to
satisfy the wishes of all who asked favours
from me; but I found, that demands upon
my compliance increased beyond the possibility of my means to gratify : a few re-
fusals gave much offence, and overbalanced
the kindness, which I had excited by favours of far greater value; so that a short

course of experience fortunately taught
me, that justice, and my own sense of generosity, should be my guides; and that
yielding to the representations of those,
who could without any compunction frame
a plausible tale, would in time lead me into
serious difficulties, and would put it out of
my power to do any thing for others, who
had just claims upon me.

Shortly after my father's death, I was
suddenly informed, that my sister was married to an officer, with whom I thought that
she had been but slightly acquainted. I
loved my sister with the greatest affection,
and I suppose my pride was wounded by
her not having acquainted me with her attachment to this gentleman. Whatever
was the cause of the emotion, which I felt
at hearing this unexpected news, its violence was extraordinary. No circumstance
or misfortune of my whole life ever affected
me suddenly with such pain and disappointment; I felt a kind of fever instantly kin-
dle in my brain; objects became confused
before my eyes; and, had not a burst of
tears relieved me, I think that some injury
to my health must have ensued. My eyes

were never used to the melting mood; but on this occasion tears relieved, settled, and satisfied my mind, in a manner that was entirely new to me. I went immediately to my sister, who regretted the pain she had given me, and whose feelings at seeing me were, I believe, nearly as acute as my own.

Her marriage turned out fortunate. Her husband* and she, mutually attached, bred up their children successfully, and have enjoyed, and I hope may long enjoy, a large portion of domestic felicity! My sister and her husband accompanied me to England, to Hare Hatch, where we lived during some months, without the occurrence of any circumstance worth relating.

Mr. Day had now returned from his first expedition to France, and had taken a pleasant house at Stow-Hill, close to Lichfield. Here he steadily pursued his plan of educating his pupil, Sabrina; and, what was something singular, all the ladies of the place kindly took notice of the girl, and attributed to Mr. Day none but the

* John Ruxton, Esq., of Black-Castle, in the county of Meath.

real motives of his conduct. The bishop's palace at Lichfield, where Mr Seward, a canon of the cathedral, resided, was the resort of every person in that neighbourhood, who had any taste for letters. Every stranger, who came well recommended to Lichfield, brought letters to the palace. This popularity in the literary world was well deserved, for Mr. Seward was a man of learning and taste; he was fond of conversation, in which he bore a considerable part, goodnatured, and indulgent to the little foibles of others: he scarcely seemed to notice any animadversions, that were made upon his own. His simplicity, or what we understand by the French word *naïveté*, was beyond what could easily be believed of a man of such talents, or of one who had seen any thing of the world. Mrs. Seward was a handsome woman, of agreeable manners, she was generous, possessed of good sense, and capable of strong affection. They had two daughters, Anna and Sally.

Anna Seward is well known to the world by her works. Miss Sally Seward had her mother's good sense, without being endowed with her father's talents. Her manners were mild, gentle, and retired: capable of much genuine feeling, she was strongly attached to her friends, and had the power of attaching others to herself.

Under the fond and truly maternal care of Mrs. Seward was bred up Miss Honora Sneyd, daughter of Edward Sneyd, Esq., youngest son of Ralph Sneyd, Esq., of Bishton, in Staffordshire. Mr. Sneyd had become a widower early in life. He had been in great affliction at the death of his wife; and his relations and friends, who were numerous, had been eager to take charge of his daughters. Nothing could exceed the kindness, and care, with which Mrs. Seward executed the trust that she had undertaken. Nobody could have distinguished Miss Honora Sneyd from Mrs. Seward's own daughters by any thing in Mrs. Seward's conduct, or in her outward deportment. Miss Sally Seward, who was nearer in age than Miss Seward to Honora Sneyd, became extremely attached to her: I have heard Honora often declare, that she felt for this lady all the tenderness of a sister; and that her own sentiments and character were formed by imitation of this

early friend of her youth. Miss Sally
Seward died when Honora was but thirteen, and she then became the more immediate pupil of Miss Seward. From
her she acquired an ardent love of literature, and she afterwards formed for herself a
refined and accurate taste. She was, however, so much eclipsed by that lady's more
shining talents, that it was not on a first
acquaintance, or to careless observers, that
Honora Sneyd's uncommon understanding
and clear judgment became conspicuous.

Soon after Mr. Day had established himself at Stow-Hill, he became intimate at
the palace. His superior abilities, lofty
sentiments, and singularity of manners,
made him appear at Lichfield as a phenomenon: his unbounded charity to the poor,
and his munificence to those of a higher
class, who were in distress, won the esteem
of all ranks; so that his breeding up a
young girl in his house, without any female
to take care of her, created no scandal,
and appeared quite natural and free from
impropriety. Sabrina, his ward, was received at the palace with tenderness and
regard. She became a link between Mr.

Day and Mr. Seward's family, that united them very strongly.

In the year 1770, I spent some time at Christmas with my friend Mr. Day, at Stow-Hill. We went every day to Lichfield, and most days to the palace, where the agreeable conversation of the whole family, and in particular the sprightliness and literary talents of Miss Seward, engaged us to pass many agreeable hours.

During this intercourse, I perceived the superiority of Miss Honora Sneyd's capacity. Her memory was not copiously stored with poetry; and though no way deficient, her knowledge had not been much enlarged by books; but her sentiments were on all subjects so just, and were delivered with such blushing modesty, (though not without an air of conscious worth,) as to command attention from every one capable of appreciating female excellence. Her person was graceful, her features beautiful, and their expression such as to heighten the eloquence of every thing she said. I was six and twenty; and now, for the first time in my life, I saw a woman that equalled the picture of perfection, which existed in my imagination. I had long suffered much from the want of that cheerfulness in a wife, without which marriage could not be agreeable to a man of such a temper as mine. I had borne this evil, I believe, with patience; but my not being happy at home exposed me to the danger of being too happy elsewhere.

The charms and superior character of Miss Honora Sneyd made an impression on my mind, such as I had never felt before. Dr. Small, about this period, settled in Birmingham; Mr. Keir was also in the neighbourhood; Dr. Darwin spent his vacant hours among us; and all these gentlemen were unanimous in their approbation of this lady.

Mr. Day alone was blind to the superiority of her character. She danced too well; she had too much an air of fashion in her dress and manners; and her arms were not sufficiently round and white to please him. However, the more I saw, the more I admired this lady. She conversed with me with freedom; and seemed to feel, that I was the first person, who had seen the full value of her character. Miss Seward shone so brightly, that all objects within her sphere were dimmed by her lustre. When she perceived the impression, that her young friend had made upon me,—an impression, which I believe she discovered, long before I had discovered it myself,—she never shewed any of that mean jealousy, which is common among young women, when they find that one of their companions, who had never before been thought equal to themselves, is suddenly treated with preeminence. On the contrary, she seemed gratified by the praises bestowed upon her friend, and took every opportunity of placing whatever was said or done by Honora in the most advantageous point of view.

Whilst I was upon this visit, Mr. André, afterwards Major André, who lost his life so unfortunately in America, came to Lichfield. This gentleman had met Miss Honora Sneyd at Buxton, or Matlock, and had paid his addresses to her, without at first meeting with any discouragement from Mrs. André, the young gentleman's mother, or, I believe, from Mr. Sneyd. But when the parents on both sides came to consider more coolly, they saw that a match, even where each party seemed well suited to the other, should not be determined upon, till a sufficient fortune for a comfortable maintenance could be provided between them. As Mrs. André and Mr. Sneyd had the good sense to raise no obstacles between their children, making only a kind and reasonable statement of the imprudence of marrying young without a sufficient fortune, no romantic difficulties urged them to resistance, and they, by tacit consent, kept at a distance from each other, except when accidental circumstances brought them together. The first time I saw Major André at the palace, I did not perceive from his manner, or from that of the young lady, that any attachment subsisted between them. On the contrary, from the great attention which Miss Seward paid to him, and from the constant admiration which Mr. André bestowed on her, I thought, that, though there was a considerable disproportion in their ages, there might exist some courtship between them. Miss Seward, however, undeceived me. I never met Mr. André again; and from all that I then saw, or have since known, I believe that

Miss Honora Sneyd was never much disappointed by the conclusion of this attachment. Mr. André appeared to me pleased and dazzled by the lady. She admired and estimated highly his talents; but he did not possess the reasoning mind, which she required. Miss Seward, in a note to her *" Monody on the Death of Major André,"* has asserted, that Mr. André, in despair upon the marriage of Honora Sneyd, quitted his business as a merchant, insinuating that he was—" Out with it !"—*jilted* by that lady,—and that, in consequence of this disappointment, he went into the army, and quitted this country.

The fact is, that Major André's first commission was dated March the 4th, 1771, and Miss Honora Sneyd was married the 17th of July, 1773—that is, two years after Mr. André went into the army. Despair, on hearing of the marriage of Honora Sneyd, could not have driven him to quit his profession, and his country; he having quitted both two years before that marriage.

To return to the course of my narrative, from which I could not help starting, to set this misstatement right—Mr. Day knew all my feelings; in the progress of my acquaintance with this lady, he heard me talk in no cold strains of Miss Honora Sneyd's perfections.—I continually expressed surprise at his not seeming to prefer her, but for some time he continued to be insensible of Honora's charms. Sabrina Sydney had, perhaps, preoccupied his mind. She was now too old to remain in my friend's house without a protectress; he therefore determined to put her to a very reputable boarding-school at Sutton-Colefield. Here it was intended, that she should improve in reading, writing, and arithmetic, and in all the useful species of accomplishments. To make a musician or a dancer of his pupil was far from his wish. During the early part of the year 1771, Mr. Day's intentions with regard to Sabrina began to change, for his mind turned toward Miss Honora Sneyd. He learned from her friend, that this lady had no engagement or attachment, that could prevent his success, if he could convince her, that the views of life, and the plan of happiness in marriage, which he had laid down, could be made compatible with those, which she had de-

termined to pursue. Few courtships ever
began between such young people with so
little appearance of romance : both however were perfectly sincere and in earnest, and for many months they were asked
together to every party at Lichfield, and
were allowed, by a kind of tacit consent,
to converse with each other, to have every
reasonable opportunity of becoming acquainted with each other's tempers, tastes,
and dispositions.

After spending some time with Mr. Day, I
returned home to Hare Hatch. My friend's
letters informed me of his feelings and opinions : I soon perceived, that he had not
only become sensible of the strength of understanding possessed by the object of his
attention ; but that, beside the value of her
mind, he began to feel the power of her
charms. He thought his suit in such a
favourable train, that he questioned me
about my feelings. He wrote me one of
the most eloquent letters that I ever read,
to point out to me the folly and meanness
of indulging a hopeless passion for any woman, let her merit be what it might; declaring, at the same time, that he " never

would marry so as to divide himself from his chosen friend." "Tell me," said he, " have you sufficient strength of mind, totally to subdue love, that cannot be indulged compatibly with peace, or honour, or virtue?"

I answered, that nothing but trial could make me acquainted with the influence, which reason might have over my feelings; that I would go with my family to Lichfield, where I should be in the company of the dangerous object; and that I would faithfully acquaint him with all my thoughts and feelings. We went to Lichfield, and staid there for some time with Mr. Day. I saw him continually in company with Honora Sneyd: I saw, that he was received with approbation, and that he looked forward to marrying her at no very distant period. When I saw this, I can affirm with truth, that I felt pleasure, and even exultation. I looked to the happiness of two people, for whom I had the most perfect esteem, without the intervention of a single sentiment or feeling, that could make me suspect I should ever repent having been instrumental to their

union. I was the depositary of every
thought, that passed in the mind of Mr.
Day; and from every thing he told me,
and from my own observations, I was persuaded, that nothing now was wanting, but
a declaration on his part, and compliance
on the part of the lady.

Just at this period, when we were walking
together one summer's evening in the Close,
a public walk at Lichfield, which was then
much frequented by the young people,
something was said in allusion to the intended match; and Miss Honora Sneyd,
in reply, expressed doubts as to its conclusion. I supposed, that she adverted to
the state of Mr. Day's mind; and I warmly gave it as my opinion, that nothing was
likely to prevent what I so much desired.
She shook her head. The next morning,
Mr. Day, in a very solemn manner, delivered to me a packet of some sheets of
paper, which he said was a proposal of
marriage to Honora Sneyd. " It contains,"
said he, "the sum of many conversations,
that have passed between us. I am satisfied," he continued, " that, if the plan of
life I have here laid down meets her ap-

probation, we shall be perfectly happy. Honora Sneyd is so reasonable, so perfect- ly sincere, and so much to be relied on, that, if once she resolves to live a calm, secluded life, she will never wish to return to more gay or splendid scenes. If she once turn away from public admiration, she will never look back again with re- gret."

I took the packet; my friend request- ed, that I would go to the palace, and deli- ver it myself. I went—and I delivered it with real satisfaction to Honora. She de- sired me to come the next morning for an answer. Mr. Day expressed extreme anxiety during the interval; more, indeed, than I had ever heard him acknowledge upon any other occasion.

In the morning I received an answer, which, from the manner in which it was delivered to me, seemed to require a farther communication. I gave it to Mr. Day, and left him to peruse it by himself. When I returned, I found him actually in a fever. The letter contained an excellent answer to his arguments in favour of the rights of men, and a clear, dispassionate view of the rights of women.

Miss Honora Sneyd would not admit the unqualified control of a husband over all her actions; she did not feel, that seclusion from society was indispensably necessary to preserve female virtue, or to secure domestic happiness. Upon terms of reasonable equality, she supposed, that mutual confidence might best subsist; she said, that, as Mr. Day had decidedly declared his determination to live in perfect seclusion from what is usually called the world, it was fit she should decidedly declare, that she would not change her present mode of life, with which she had no reason to be dissatisfied, for any dark and untried system, that could be proposed to her.

Mr. Day continued really ill for some days. Dr. Darwin ordered him to be bled, and administered, wisely, to that part of him which was most diseased—his mind. The intimacy, which subsisted among the inhabitants of Lichfield, prevented any estrangement between Mr. Day and the family at the palace; and in some weeks a new ob-

ject of attention was presented to the Lichfield world. Mr. Sneyd (Honora's father),
who had hitherto lived in London, now
came to establish himself at Lichfield. He
assembled all his daughters to reside with
him, and with them came his fifth daughter, Miss Elizabeth Sneyd. She had lived
with Mr. and Mrs. Powys (of the Abbey,
Shrewsbury), the kind friends and relations
by whom she had been educated.—Mrs.
Powys, wife to Henry Powys, Esq. was
Mr. Sneyd's niece.

Circumstances, in themselves entirely
trivial, become to certain persons in peculiar situations highly interesting. In relating the events of life, these trivial circumstances force themselves upon the attention, and obtain, as it were, by impertinent intrusion, a place in the narrative.
To balance this inconvenience, these very
circumstances give an air of reality, and
of that perfect simplicity, which ought to
be apparent in every page of biography
This reflexion occurred, when I began to
mention the first appearance of Miss Elizabeth Sneyd at Lichfield, which, though
quite uninteresting to me at the time, was

marked by the vivid interest awakened in my mind at this period of my life.

I had introduced archery as an amusement among the gentlemen in the neighbourhood; and had proposed a prize of a silver arrow, to be shot for at a bowling-green, where our butts had been erected. All the ladies, who frequented the amusements of Lichfield, were assembled, and Miss Seward appeared with her usual sprightliness and address, accompanied by Honora.

We had musick and dancing; some of the gentlemen fenced, and vaulted, and leaped; and the summer's evening was spent with as much innocent cheerfulness, as any evening that I can remember. Miss Elizabeth Sneyd and her father came among us in the middle of our amusements; just as a country dance was nearly ended, Miss Honora Sneyd introduced me to her sister, desiring me to dance with her, to prevent her being engaged by some stranger, with whom they might afterwards not chuse to form an acquaintance. Miss Elizabeth Sneyd was, in the opinion of half the persons who knew them, the hand-

somest of the two sisters; her eyes were uncommonly beautiful and expressive, she
was of a clear brown, and of a more healthy
complexion than Honora. She had acquired more literature, had more what is
called the manners of a person of fashion,
had more wit, more vivacity, and certainly
more humour than her sister. She had,
however, less personal grace; she walked
heavily, danced indifferently, had much
less energy of manner and of character, and
was not endowed with, or had not then acquired, the same powers of reasoning, the
same inquiring range of understanding, the
same love of science, or, in one word, the
same decisive judgment as her sister.

Notwithstanding something fashionable
in this young lady's appearance, Mr. Day
observed her with complacent attention.
Her dancing but indifferently, and with no
symptom of delight, pleased Mr. Day's
fancy; her conversation was playful, and
never disputatious, so that Mr. Day had liberty and room enough, to descant at large
and at length upon whatever became the
subject of conversation. She was struck
by his eloquence, her attention was awak-

ened by the novelty of his opinions; he appeared to her young mind as the most extraordinary and romantic person in the world. His educating a young girl for his wife, his unbounded generosity, his scorn of wealth and titles, his romantic notions of love, which led him to think, that, where it was mutual and genuine, the rest of the world vanished, and lovers became all in all to each other, made a deep impression upon her, and made her imagine, that, if such a man loved her with truth and violence, she was capable of as strong attachment, and of as disinterested sacrifices, as could be made to her. Every body perceived, that Miss Elizabeth Sneyd had made a greater impression in three weeks upon Mr. Day, than her superior sister had made in twelve months.

One restraint, which had acted long and steadily upon my feelings, was now removed: my friend was no longer attached to Miss Honora Sneyd. My former admiration of her returned with unabated ardour. The more I compared her with other women, the more I was obliged to acknowledge her superiority. This admiration was

unknown to every body but Mr. Day. He
could not see more plainly than I did the
imprudence and folly of becoming too fond
of an object, which I could not hope to obtain. With all the eloquence of virtue and
of friendship, he represented to me the danger, the criminality of such an attachment.
I knew, that there is but one certain method
of escaping such dangers—*flight*.

I resolved to go abroad : Mr. Day determined to accompany me to France, and
to dedicate a large portion of his time to
the acquirement of those accomplishments,
which he had formerly treated with sovereign contempt. Miss Elizabeth Sneyd had
convinced him, that he could not with propriety abuse and ridicule talents, in which
he appeared obviously deficient. She told
him, that she considered such acquirements
as frivolous, in many situations ridiculous ;
but that she could not be satisfied with
the abhorrence, which upon all occasions
he expressed, of accomplishments, which
he had not been able to attain. On her
part she promised not to go to London,
Bath, or any other public place of amusement, till his return ; and she engaged with

alacrity, to prosecute an excellent course of reading, which they had agreed upon before his departure.

As to myself, I took the most scrupulous pains to avoid whatever might induce any of our acquaintance to think, that I felt more than common esteem for Miss Honora Sneyd. I took every opportunity of declaring my intention of settling in Ireland, whenever I should return from France; and, in various incidental conversations, I endeavoured to convince her, that young women, who had not large fortunes, should not disdain to marry, even though the romantic notions of finding heroes, or prodigies of men, might not be entirely gratified. Honora listened, and assented; and I left England with perfect conviction, that I had not endangered the happiness of any of my friends, and that I had not devoted myself to unavailing regret, or unreasonable hope.

CHAPTER X.

———

MR. DAY and I quitted England, and we took with us my son, who was then about seven or eight years old. I engaged a gentleman to travel with us as my boy's tutor, so that we might be at liberty to go where we pleased, without being under anxiety for our pupil. I hoped to teach the child to speak French with a proper accent, and without his being obliged to learn it as a dead language: and I at the same time thought, that I should secure him from the danger of being spoiled during my absence, if I left him behind me. I am not sure, that I judged rightly; but, if I now doubt, it arises perhaps from my perceiving the inconveniences which ensued from my having made this decision. I am not able to determine, what might have been the result of a different conduct.

I must not here omit a remarkable circumstance, which ought to be recorded in justice to Rousseau's penetration in judging of children. In passing through Paris at this time, we went to see him : he took a good deal of notice of my boy; I asked him to tell me any thing that struck him in the child's manners or conversation. He took my son with him in his usual morning's walk, and when he came back, Rousseau told me, that, as far as he could judge from two hours observation, he thought him a boy of abilities, which had been well cultivated; and that in particular his answers to some questions on history proved, contrary to the opinion given in Emilius and Sophia, that history can be advantageously learned by children, if it be taught reasonably, and not merely by rote. " But," said Rousseau, " I remark in your son a propensity to party prejudice, which will be a great blemish in his character."

I asked how he could in so short a time form so decided an opinion. He told me, that, whenever my son saw a handsome horse, or a handsome carriage in the street, he always exclaimed, "That is an English horse, or an English carriage!" And that, even down to a pair of shoe-buckles, every thing that appeared to be good of its kind was always pronounced by him to be English. "This sort of party prejudice," said Rousseau, "if suffered to become a ruling motive in his mind, will lead to a thousand evils: for not only will his own country, his own village, or club, or even a knot of his private acquaintance, be the object of his exclusive admiration; but he will be governed by his companions, whatever they may be, and they will become the arbiters of his destiny."

In fact, the boy had the species of party spirit, which Rousseau remarked, and this prophecy, as after events proved, shewed his sagacity.

We remained but two days at Paris, for Mr. Day had very little curiosity as to the common objects, which engage the attention of travellers. My son was too young, to learn any thing of real advantage by visiting public monuments, or libraries, or galleries of pictures; and at that time I was more intent upon seeing varieties of people, than the works of art; in this plan I was most certainly mistaken. I might have acquired a very large portion of useful and ornamental knowledge, had I passed a few weeks at Paris.

We proceeded to Lyons, like true English travellers, without stopping on the road to examine what was curious, or worthy of observation. We determined to pass the winter at Lyons, as it was a place where excellent masters of all sorts were to be found; and here Mr. Day put himself to every species of torture, ordinary and extraordinary, to compel his antigallican limbs, in spite of their natural rigidity, to dance, and fence, and manage the *great horse*. To perform his promise to Miss E. Sneyd honorably, he gave up seven or eight hours of the day to these exercises, for which he had not the slightest taste, and for which, except horsemanship, he manifested the most sovereign contempt. It was astonishing to behold the energy, with which he persevered in these pursuits. I have seen him stand between two boards, which reached from the ground higher than his knees : these boards were adjusted with screws, so as barely to permit him to bend his knees, and to rise up and sink down. By these means M. Huise proposed to force Mr. Day's knees outward; but his screwing was in vain. He succeeded in torturing his patient ; but original formation, and inveterate habit, resisted all his endeavours at personal improvement. I could not help pitying my philosophic friend, pent up in durance vile for hours together, with his feet in the stocks, a book in his hand, and contempt in his heart.

In the mean time I lodged myself in excellent and cheerful apartments upon the ramparts. I boarded in the family of a gentleman, who was at the head of the Military Academy at Lyons, where I soon learned to speak French, so as to be intelligible. I read and understood it well; and I systematically avoided any attempt to write that language with critical accuracy, as I believed, and still believe, that in composition or eloquence " one language only can one genius fit;" and that, however the literature of various countries may enrich the mind, a mixture of their vocabularies in writing, or speaking, is always prejudicial. The curse of Babel lights upon most of those, who speak a leash of languages at once. I do not mean to say, that an accurate knowledge of the written language of France, Germany, and Italy, is hurtful : on the contrary, I am sensible of the advantages, that may be obtained by a critical knowledge of any of them, and of the inexhaustible entertainment which they afford; what I object to is, such a habit of writing or speaking them, as induces the mind to *think* partly in one language, and partly in another. For instance, I acquired at Lyons such a knowledge of the technical vocabulary of mechanics, that I was able to express myself clearly on such subjects in French, though I could not explain myself on them in English; and I have at this instant by me a treatise on watermills, which I wrote in French, and which I never could induce myself to translate, because it was difficult to me to find appropriate terms in my own language.

Monsieur Charpentier, who was the master of the Academy at Lyons, had seen much of the world, and communicated agreeably what he had seen. He had been controller of the household to the embassy at Constantinople for upwards of twenty years, and had been no inattentive observer. I remember distinctly his having mentioned to me many things relative to Turkey, which I afterwards read in de Tott's Memoirs; and which I heard condemned in de Tott as travellers' wonders. Madame Charpentier was young, beautiful, lively, and accomplished, of an excellent disposition, and less fond of publick amusements than most French women. During nearly two years that I was at Lyons, I never had occasion to repent my having established myself in her family, as I met with uniform kindness and confidence from every part of it. I had letters of introduction from several quarters, which made me soon acquainted with the houses, where strangers were received. Monsieur de Verpillier (or perhaps Viripillier) was at that time Commandant of Lyons; his son was Town-Major; hospitable, gallant, dissipated, and very fond of the English.

To continue in this society, which was of course considered as the court of Lyons, it was necessary to play—and for play I had little taste, and much salutary horror. I therefore resolved to lay aside a certain sum, I think a hundred and fifty louis d'ors, to play with, and these *I determined to lose.* I began with caution, took pains to learn the game, and played with so much attention, that I was considered as a young man whose secret passion was gaming. I was permitted to win, and, by degrees, as fortune favoured me, I learned how to win, when I was not permitted. The play was in general far from *desperate;* but whenever any body sallied beyond the ordinary course, I always took him up, and was so frequently successful, that I did not often meet with competitors. In short, I played with so much boldness, and was so much at my ease, that I had all the advantages that I could possibly expect; and as winning or losing was perfectly indifferent to me, I had an opportunity of observing manners and character at leisure, while the minds of the company were engaged or inflamed by the course of the game. My destined sum did not fail, till long after my having obtained by it all that I desired. When the last louis d'or had left me, I quitted the field with a joyful heart, and retired to live with such society as I had selected from my usual associates. I continued to visit these, but I never was for a moment tempted again to play, while I remained at Lyons.

An occupation far more agreeable to my taste soon presented itself to me. A project for enlarging the city of Lyons had been formed by Monsieur Perache, an architect of that city. A company had subscribed a large sum to carry it on; and it was considered as a speculation, that promised much advantage to the city, and much profit to the company. Walking one morning with a gentleman, who was deeply concerned in this project, I adverted to some obvious mistakes, that appeared to me to have been made in the progress of the work; and as he perceived, that I knew something of engineering, he invited me to meet Perache and some of the principal associates. We met: Pe-

rache listened with temper and attention
to what I said ; and it was proposed, that
I should conduct a certain part of the
work for some time. In this I succeeded
so well, that I was pressed to manage a
large part of the business, which required
skill and boldness. I undertook it upon a
singular stipulation, that I should draw for
a certain sum of money on the treasurer
of the company every Saturday, without
giving any other account than a simple
statement under my hand, that I had laid it
out to the best of my ability for the benefit
of the company.

The object of the scheme was to turn
the course of the Rhone from its usual
channel, which hemmed in the city on the
only side, toward which it could be extended. Ancient historians describe Lyons, Lug-
dunum, in the time of Claudius, as standing on a confined spot of ground, resembling in shape the Greek letter Delta. A
description which had continued to be
exact from that time, till the period when
this project was undertaken. The Saone
on one side, steep hills on the other, and
the Rhone rushing forward with great vio-

lence at the bottom of these hills, and turning nearly at right angles from its direct course, to meet the Saone beneath the ramparts, shut in the city, and prevented the extension of its buildings. Now to enlarge the town, it was proposed to divert the Rhone into a new channel, so as to form a junction with the Saone nearly a mile farther from the town than it was at that time. The deserted bed of the river was to be filled with the rubbish of the city, so as to obtain ground for extending the buildings in that quarter. After a new channel had been partly cut for the river, it was proposed to turn the rapid course of the Rhone into this new channel, with the hope, that, as the soil was light and sandy, the violence of the flood, from the melting of snow in the distant mountains of Savoy, would excavate a bed of sufficient capacity to receive the whole of the river without any farther trouble. By piles and earth, an embankment had been effected, that nearly shut out the Rhone from the Saone. A gap of not more than twenty or thirty feet remained to be closed. To expedite this work, I constructed the following apparatus, instead of using the mere force of hands, which had hitherto been alone employed.

To cross the Rhone, a species of ferry-boat was in use, which required no labour in the passage. It was a large strong boat, with sides nearly parallel; it was held from going with the current by a rope of twenty or thirty yards long: this rope, by means of a large pulley, was attached to a strong cable, that was stretched across the river, from posts fifteen or eighteen feet high, that were elevated on each bank. The rope that held this boat was, by these means, kept nearly in the direction of the stream, and at right angles to the cable, which was stretched across the river, so that as the boat moved, this rope ran along the cable, and did not retard its progress. To move the boat, no force but that of the stream was employed. The rope, by which it was held, was hooked upon a staple, fastened near one end of one side of the boat, that is to say, at about one third, or one fourth of the length of the boat, from the end. When the boat was pushed off from the land, the stream forced the boat into an oblique direction, because of the situation of the rope. The current was thus obliged to act upon the boat as upon an inclined plane; so that, being prevented from moving in the direction of the current, the boat was forced in the only direction in which it could move, that is, across the river. These boats were called *trailles*. To transport earth and gravel into the opening which was to be filled up, men had been employed to carry baskets of earth on their heads into the boat, and, after the boat had crossed, to empty them into the water. A great number of men were thus employed, who could effect but little, as they were in each others' way in getting into the boat, and still more so in emptying their baskets.

To remedy this inconvenience, I had two strong piles erected, twelve feet above the surface of the water, on the side where the earth was dug; upon these there was placed a moving platform, like a very large cart with low sides, the whole turning on an axle-tree, supported in a fork on the top of each of the piles. This platform was sixteen or eighteen feet square, and could contain a large quantity of gravel. A gangway, or inclined scaffolding of boards, led to this cart from the ground, by which labourers with baskets could ascend and fill it. A similar platform was erected on the *traille*, or boat, into which the contents of that which was filled on the shore were emptied. The boat then passed the river, and the load was emptied at once into the water. In the meantime, the labourers filled the land platform, which was ready to be emptied into the *traille* at the moment of its return. This contrivance succeeded perfectly well: not a moment was wasted, and the whole contents of the platform belonging to the boat could be emptied exactly where they were wanted.

I also constructed, across a *ravine*, a bridge, supported on slight trestles, that was sufficiently strong and broad to hold a wheelbarrow. On this bridge, which was inclined downwards to the place where the barrows were to be emptied, by means of a cradle drawn up by pullies, I mounted a number of wheelbarrows, which followed each other, and, from the manner in which they were guided, reached their destination in regular succession. A bridge, wide enough to receive the labourers, as well as their loads, would have been preferable in point of expedition; but I could not procure a sufficient number of men willing, or indeed able, to pass over a gangway fifty or sixty yards long, and sixteen or seventeen and in some places twenty feet from the ground. A floor of four feet wide would not have been sufficient, unless it were railed in; an expense, which would have been too great for the purpose. This gangway was · only the breadth of one board; and, as it was a new and entertaining sight to see the wheelbarrows pass over it without human guides, crowds of people came from Lyons to look at the spectacle. To gain a few pence by his boldness, one of the labourers used to commit himself to the uncertain aerial path; but unfortunately one day, when Madame l'Intendante of Lyons appeared, he, to outdo his usual outdoings, performed some gambols, which overturned his wheelbarrow and broke his arm.

I immediately took down this apparatus,

and substituted a less dangerous system, which, though more costly, would be free from hazard. I was satisfied, that this instance of prudence gained me friends; because those who approved of it were sensible, that it must cost me some struggle to give up before numbers a plan, which had succeeded as to the object for which it was intended, and which had novelty, and perhaps some ingenuity, to recommend it. No other accident happened, and the work went on rapidly and prosperously.

As soon as I had engaged in this work, and found that I should be likely to continue some time at Lyons, I invited my wife to come over to me; I had left her with her father and sisters at Black-Bourton. Accompanied by one of her sisters, she arrived at Lyons, where she staid till winter; at the commencement of which, she being weary of French society, and anxious to be in England before the time when she expected to be confined, as she had a dread of lying in at Lyons, or in any part of France, took an opportunity of returning to England, under the care of my friend, Mr. Day. He having now learnt all that it was possible for riding, fencing, and dancing-masters to teach him, determined to go back to England, to claim, as the reward of his labours, the hand of Miss Elizabeth Sneyd.

During the time when I was engaged in the conduct of the public works at Lyons, I had not leisure to attend sufficiently to my son. While my friend, Mr. Day, remained with me, he was so kind as to pay particular attention to him ; but after Mr. Day left us, my boy was under the care of a tutor, whom I had brought from England. My son was then almost nine years old ; he had considerable abilities, uncommon strength and hardiness of body, great vivacity, and was not a little disposed to think and act for himself. I had begun his education upon the mistaken principles of Rousseau ; and I had pursued them with as much steadiness, and, so far as they could be advantageous, with as much success as I could desire. Whatever regarded the health, strength, and agility of my son, had amply justified the system of my master ; but I found myself entangled in difficulties with regard to my child's mind and temper.

He was generous, brave, good-natured, and
what is commonly called good-tempered;
but he was scarcely to be controlled. It
was difficult to urge him to any thing that
did not suit his fancy, and more difficult
to restrain him from what he wished to follow. In short, he was self-willed, from a
spirit of independence, which had been in-
culcated by his early education, and which
he cherished the more from the inexperi-
ence of his own powers.

I must here acknowledge, with deep regret, not only the error of a theory,
which I had adopted at a very early age,
when older and wiser persons than myself
had been dazzled by the eloquence of
Rousseau; but I must also reproach myself with not having, after my arrival in
France, paid as much attention to my boy as
I had done in England, or as much as was
necessary to prevent the formation of those
habits, which could never afterwards· be
eradicated. I dwell on this painful subject, to warn other parents against the er-
rors, which I committed. I had successfully reached a certain point in the education of my pupil; I had acquired complete

ascendancy over his mind; he respected
and loved me; but, relying upon what I
had already done, I trusted him to the care
of another, who, with the best intentions
in the world, had no experience in the
management of children, or any habitual
influence over his particular pupil. The
boy soon obtained the mastery. The tutor
was a man of abilities, and truly solicitous
to discharge his duty; but he was of an
easy temper, and his mind was intent upon
objects of his own. He had a slight impediment in his speech, and had not a favourable disposition for learning languages.
He had a French master, to whom he dedicated at least two hours every day. My
son was invited, and tempted by various
means, to partake of the lessons, to which
his tutor so assiduously attended; but the
boy could never be induced to get by rote
the French irregular verbs, or to hear critical remarks upon the uses of certain
common particles, which strangers are apt
to confound and misapply. But in the
mean time he learned to speak French
fluently, and with a good accent; and before his tutor could express his wants at

dinner with common accuracy, or indeed before he became intelligible to the people with whom he lived, my son was able to read and converse without any hesitation. The consequence might be easily foreseen. The boy perceived his superiority upon a subject of mutual pursuit; and the tutor, who had himself failed in learning French, could never afterwards persuade his pupil to learn Latin in the usual dull routine; neither could he induce him to apply steadily to any species of study, that required sedentary habits, or continued attention.

Even then, had I perceived, or, to speak with more candour, had I vigorously counteracted these faults, they might have been cured. But I had entered into engagements with the company who were employed in turning the course of the Rhone, which prevented me from applying myself to the education of my son. I therefore determined to send him to school, during the time that I should remain at Lyons.

The Jesuits had lately been dismissed from their magnificent establishment in that city, and *les pères de l'oratoire* had been introduced in their place. The neatness order, extent, and grandeur of the building, the regulations of the seminary, and the deportment and high character of its masters, had prepossessed me much in its favour.

One point of considerable moment, however made me hesitate. There were great hazards in sending a boy to a Catholic seminary, at an age when he was liable to receive indelible prejudices; and in committing him to the care of persons, who thought the attempt at conversion a merit, and who believed it to be a duty, as the only means of saving a heretic soul from perdition. I went to the Father, who was then at the head of the institution, and stated to him with perfect openness my views and apprehensions.

The old man, in whose countenance benevolence and wisdom were strongly portrayed, assured me, that he would do every thing in his power to prevent any interference in the religious principles of my child; he said, that he deplored the blindness of my countrymen; but that, as to me, he would receive the charge of my son as a trust reposed in him by me, which he was bound to execute according to my wishes, when he agreed to accept it. He added, that he knew how I was employed at Lyons, and he perceived how that employment interfered with the education of my son; he was therefore willing to have a share in cultivating talents, which he heard highly spoken of among his friends. In short he promised not to meddle with my son's religion, and to inform me, if he found that any of his under-masters disobeyed his orders on this subject.

About a month afterwards I paid a visit to the reverend father; he told me, that notwithstanding his injunctions to the contrary, one of the under-masters had endeavoured to teach my son such doctrines, as he thought necessary for his salvation. " I will tell you," said the father, " exactly what passed: Le père Jerome, from the time your son came, had formed the pious design of converting your little gentleman; and for this purpose he had taken particular notice of him, and had from time to time given him bonbons. One day he took your boy between his knees, and began from the beginning of things to teach him what he ought to believe. 'My little man,' said he, 'Did you ever hear of God?'

" 'Yes.'

" 'You know, that, before he made the world, his spirit brooded over the vast deep, which was a great sea without shores, and *without bottom*. Then he made this world out of earth.'

" 'Where did he find the *earth?*' asked the boy.

" 'At the bottom of the sea,' replied father Jerome.

" 'But,' said the boy, 'you told me just now, that the sea had no bottom!' "

The Superior of the collége des oratoires concluded, " You may, Sir, I think, be secure, that your son, when capable of making such a reply, is in no great danger of becoming a catholic from the lectures of such profound teachers as these."

The Superior kept his word with me, and I never had reason to believe, that any farther attempts at conversion were made upon my son.

CHAPTER XI.

———

ON every change of residence, in every place to which we go, we probably find some individual more suited to our taste than any other. During my residence at Lyons, I felt this sort of predilection for the Marquis de la Poype, a gentleman with whom I then first became acquainted. He was a man of much general information, well acquainted with English literature, and sufficiently master of the language, to be able to hold a conversation upon common subjects. His manners were peculiarly pleasing, and he was married to a beautiful woman, whom he loved with an affection, that was not often known to subsist between married people in France, or at least in the part of

France which I had seen. He pressed me
much to pay him a visit at his Chateau in
Dauphiny; and at length I promised to
pass with him some of the Christmas holidays. An English gentleman of the name
of Rosenhagen went with me.

We arrived in the evening at a very
antique building, surrounded by a moat,
and with gardens laid out in the style
which was common in England in the beginning of the last century. These were
enclosed by high walls, intersected by canals, and cut into parterres by sandy walks.
We were ushered into a good drawing
room, the walls of which were furnished
with ancient tapestry. When dinner was
served, we crossed a large and lofty hall,
that was hung round with armour, and
with the spoils of the chace; we passed into
a moderate sized eating room, in which
there was no visible fireplace, but which
was sufficiently heated by invisible stoves.
The want of the cheerful light of a fire cast
a gloom over our repast, and the howling
of the wind did not contribute to lessen this
dismal effect. But the dinner was good,
and the wine, which was produced from the

vineyard close to the house, was excellent. Madame de la Poype, and two or three of her friends, who were on a visit at her house, conversed agreeably, and all feeling of winter and seclusion was forgotten.

At night, when I was shewn into my chamber, the footman asked if I chose to have my bed warmed. I inquired whether it was well aired; he assured me, with a tone of integrity, that I had nothing to fear, for " that it had not been slept in for half a year." The French are not afraid of damp beds, but they have great dread of catching some infectious disease from sleeping in any bed, in which a stranger may have recently lain.

My bed-chamber at this Chateau was hung with tapestry, and as the footman assured me of the safety of my bed, he drew aside a piece of the tapestry, which discovered a small recess in the wall, that held a *grabat*, in which my servant was invited to repose. My servant was an Englishman, whose indignation nothing but want of words to express it could have concealed : he deplored my unhappy lot ; as for himself, he declared, with a look of

horror, that nothing could induce him to
go into such a pigeon-hole. I went to visit the accommodations of my companion,
Mr. Rosenhagen; I found him in a spacious apartment hung all round with tapestry, so that there was no appearance of any
windows. I was far from being indifferent
to the comfort of a good dry bed; but
poor Mr. Rosenhagen, besides being delicate, was hypochondriac. With one of the
most rueful countenances I ever beheld,
he informed me, that he must certainly *die*
of cold; his teeth chattered whilst he
pointed to the tapestry at one end of the
room, which waved to and fro with the
wind; and, looking behind it, I found a
large stone casement window without a
single pane of glass, or shutters of any kind.
He determined not to take off his clothes;
but I, gaining courage from despair, un-
dressed, went to bed, and never slept better in my life, or ever wakened in better
health or spirits, than at ten o'clock the
next morning.

How Rosenhagen fared, or what happened to him afterwards, I really forget.
After breakfast the Marquis took us to visit
the Grotto de la Baume, which was at the

distance of not more than two leagues from his house.

We were most hospitably received at the house of an old officer, who was *Seigneur* of the place. His hall was more amply furnished with implements of the chase and spoils of the field, than any which I have ever seen, or ever heard described. There were nets of such dimensions, and of such strength, as were quite new to me: bows, cross-bows, of prodigious power: guns of a length and weight, that could not be wielded by the strength of modern arms; some with old matchlocks, and with rests to be stuck into the ground, and others with wheel-locks; besides modern fire-arms of all descriptions: horns of deer, and tusks of wild-boars, were placed in compartments in such numbers, that every part of the walls was covered either with arms or trophies.

The master of the mansion, in bulk, dress, and general appearance, was suited to the style of life, which might be expected from what we had seen at our entrance. He was above six feet high, strong, and robust, though upwards of sixty years of age; he wore a leathern jerkin, and instead of having his hair powdered, and tied in a long queue, according to the fashion of the day, he wore his own short grey locks; his address was plain, frank, and hearty; but by no means coarse or vulgar. He was of an ancient family, but of a moderate fortune. I forget whether his wife was living, but she did not appear; his son assisted him in doing the honors of the house. We were invited to dinner, to which we were to return after our visit to the grotto.

We were provided with guides and torches, and at no great distance from the castle we entered the cavern. The arch, which hung over the entrance, was not so large as that of Peake's hole, in Derbyshire; it was however more striking, because it was free from all signs of human habitation. A clear, but not a broad stream, issued through the opening, leaving a narrow dry path on each side of it, which led one or two hundred yards into the hill that covered the cavern. Here the rock dipped, in such a manner, as nearly to close the channel of the stream; but, by creeping near the ground, we got through this narrow passage, and found ourselves in another reach of the cavern, the sides of which, of solid and continued rock, lay in strata, so as to form steps of twenty or thirty inches high, to a height of fifteen or twenty feet. These were often sloping, and always slippery, so that it was a business of no common danger to scramble upwards, and a subject of terror to think of coming down again.

It is seven and thirty years since I was in Dauphiny, and I now write without an exact recollection of the sizes or distances of objects; but the objects themselves must have been striking, since they left such a durable impression on my mind. I know that there have been published accounts of the Grotto de la Baume; but I now write without reading them, purposely to try how far my memory is or is not exact; and to shew how far objects, that in youth forcibly struck the imagination, remain fixed in the mind in advanced age.

Of the chambers into which we clambered, in this cavern, I have a distinct recollection, especially of *the chamber of the bats;* into this we crept by a low and narrow opening. It is a room not less than twenty feet square, and about thirteen feet high. The dung of the bats was two or three inches in depth under our feet. This had all accumulated in the course of a few years. Three or four years previous to the time when I saw it, the chamber had been completely cleansed for the value of the manure. The bats were suspended from the roof, one below another, clustering like bees; they were then five deep, so that fifty thousand would be a moderate estimate of their numbers; and I had no reason to doubt what the old baron told me, that several acres of ground had been most highly manured with their dung, which had been collected from different parts of the cave.

While we were examining this chamber, some of the bats were disturbed; they flew about with such violence, as to extinguish several of our lights; and, had not our conductor prudently ordered one of our attendants to hurry out of the apartment before his torch was blown out, we should have been left without any light, at the mercy of these odious animals, whose stench was uncommonly noisome. We were all heartily glad to escape from their buffeting, and after quitting their den, and walking along a narrow ledge, we entered another room larger than the chamber of the bats ;—a grotto which struck me as the most brilliant and magnificent spectacle I had ever beheld. It was surrounded with exquisite sparkling incrustations, which, towards the bottom of the walls, spread upon the floor in such a manner as to form steps. From the ceiling several large stalactites depended in regular order, and in the middle of the grotto there stood a circular altar of crystal finishing in a tabular form slightly convex. The shape of this altar, or *pedestal*, as perhaps it might more properly be called, swelled gradually as it descended, so as to form a base in the *shape* of a trumpet. This, if I recollect rightly, stood upon a circular platform, which was raised on three or four steps, all of the same crystalized materials. The surface of the pedestal was formed of leaves, or laminæ, that fell over each other like the covering of a pine-apple ; but increasing in size towards the bottom, as they approached the trumpet-formed base. The steps were horizontal, not of one continued surface, but appeared like rows of large scollop shells, very shallow, about eight or ten inches diameter, and encrusted in such a manner as to resemble the coating of a rock melon. All this tracing was wonderfully regular, as are all the operations of nature where they are undisturbed, and here there was nothing to disturb her. Not a breath of air, not any motion or noise, was perceptible in the chamber, but *minute drops* from the stalactites, which hung from the ceiling. Immediately above the pedestal, which I have but inadequately described, there descends from the centre of the roof an immense stalactite, trumpet-shaped, sixteen inches in diameter at the larger end, which was attached to the ceiling, and tapering in a beautiful curve nearly to a point. The brilliancy of the gems, which sparkled on every side of us, formed a contrast the most pleasing with the stone-colored altar in the middle of the chamber.

I was so fixed in admiration of what I saw, that I had not time to examine into the philosophy of the phenomenon. That the water oozed through the rock was apparent; and that it flowed more copiously from the centre, than from any other part, was also evident; but how the surplus water was disposed of, I have no recollection.

We had taken off our coats at our entrance into the cave, yet found the heat extremely oppressive. Our guides insisted upon our returning as slowly as possible, that we might be the better able to bear the contrast of the external air, when we should leave the cavern. Notwithstanding our precautions, the open air, which was probably at forty-five degrees of heat, appeared to us like cold water. We covered ourselves as warmly as we could, and returned with our host, the old officer, to dine with him at his castle.

Our dinner was in its arrangement totally unlike any thing I had seen in France, or any where else. It consisted of a monstrous but excellent wild boar ham; this, and a large savoury pie of different sorts of game, were the principal dishes; which, with some common vegetables, amply satisfied our hunger. A good dessert, and wine of the growth of the neighbouring vineyards, crowned our repast. The blunt hospitality

of this rural baron was totally different
from that, which is to be met with in remote parts of the country of England. It
was more the open-heartedness of a soldier,
than the roughness of a squire. He entertained us with the account of rural sports,
to which we were utter strangers.

Hunting the wild boar is a dangerous
sport, and partakes more of a warlike exercise, than of a civil amusement. The
wild boar, though in general of a small size
in this part of the continent, is still fierce ;
and sometimes, though rarely, an animal of
this species, that has grown old and bulky,
by having lain hid for a length of time in
the forest, is roused from its lair, and opposes
the dogs and hunters with formidable resolution. These boars tear up the dogs, as
they run past them, with a sudden and oblique motion of their projecting tusks ; and
frequently wound the hunters themselves,
if they cannot elude their course. It seems
extraordinary, that such wild animals should
still be found in a country so highly civi-
lized.as France.

We returned by moonlight to the Chateau de la Poype well pleased with our ex-

cursion. The next day we passed in very different society. We accompanied the Marquis and his lady to the residence of a president of Grenoble; the manners here were different from what I had seen at the Chateau de la Poype, and at the houses which I frequented in Lyons. There was something formal, yet condescending, in the men; and in the ladies that mixture of pretension, pedantry, and affectation, which the French express by the word *precieuse*.

A numerous company was assembled. Among the guests were some mere country gentlemen. These gentlemen were mostly unmarried. Their rusticity was strongly marked, yet, nevertheless, I could perceive, that they were admitted to the rank of country lovers by some of the ladies. This gave them an air of pretension that was disgusting, and in a high degree ridiculous.

Before dinner I received a message from the mistress of the house, who had not appeared with the rest of the family, inviting me to her apartment. My friend, the Marchioness de la Poype, informed me, that this lady, by an unfortunate accident, had been for some years deprived of the use of

her limbs; and that, to divert her attention from her sufferings, she had applied
herself to literature with unremitting assiduity. From the representations of M. de
la Poype, she had become desirous to see
an English Gentleman, who could converse
with her on subjects suited to her taste.
I was introduced by Madame de la Poype.
In an elegant chamber, ornamented with
pictures, and furnished with books, I saw
a most beautiful creature of seven or eight
and twenty. She was supported by cush-
ions on a *chaise-longue*, and, were it not
for a smile, and the motion of her lips, she
might have passed for a figure formed in
wax. I saw, though I did not hear, that she
spoke; but when I seated myself, as I was
desired, on a low seat near her, I found that
her voice, though scarcely audible, was ar-
ticulate and melodious. She did not move
a limb; and in her countenance there
seemed to be a constant suppression of the
sense of pain. She conversed with animation upon several subjects of literature, and
various branches of knowledge, like one
who had not, for a long time, had an opportunity of communicating her thoughts to

any person engaged in pursuits similar to
her own. When dinner was announced,
she seemed as if she thought it was too
early; and as I departed, she requested,
with much timidity, that I would *give up*
another hour in the evening. As soon after
dinner as I saw the company securely
seated at the card table, I returned to the interesting invalide, with whom I passed some
hours most agreeably. She seemed more
free from prejudice and party, than almost
any person I have ever met with; fond of
reasoning, yet averse from disputation;
versed in literature, yet not ambitious to
display it. She understood English, and
was well acquainted with our best authors,
many of whom she appreciated and criti-
cised with much candour and sagacity. I
could not help deploring the situation of
this superior woman. She made no complaint, though it was obvious, that none of
the family in which she lived had tastes,
or knowledge, or sympathy, that could al-
leviate her sufferings.

This scene made a great impression upon
me; an impression that was increased by
the recollection of my poor mother, who

had been deprived of the use of her limbs at my birth. I was accustomed, in my childhood, to see her cheerful, and sometimes happy; and now I saw this lady forget, or seem to forget, pain and disease, while her mind was occupied with literature. Thus my attention to the acquisition of literature, and knowledge of every kind, as the means of happiness, independently of their being a distinction, was increased; and I determined on steadily persevering in the cultivation of my understanding.

Upon my return to Lyons these impressions were strengthened still more by female influence; especially by my becoming acquainted with Mrs. Pitt, afterwards Lady Rivers. She was then far from young, but she was uncommonly agreeable, though so deaf, that, except in a carriage, she could hardly hear what was said at the distance of a yard. I used to drive out with her, evening after evening, on the beautiful banks of the Saone, toward l'Isle de Barbe; and from her various and animated conversation I learned more of the *interior* of the great world of that day, more anecdotes of distinguished persons, and more knowledge of the human heart, than I could have obtained from any other source within my reach.

The society at Lyons was at this time emulating the polish of Parisian manners, and approaching fast to the dissipation and relaxation of morals, which prevailed in Paris. Among the trifling anecdotes, that have remained in my memory, I may mention a repartee of a belle at Lyons, a Madame *Bobu.* This lady had given some offence to M. de Verpillier, the major of Lyons. At a masquerade, the major discovered this lady in her disguise, and accosted her in a sarcastic tone, with a quotation from the syllables of the Primer;— " Comment vous portez vous, Madame Ba-Be-Bi-*Bo-Bu?*"—She answered, " Tres bien ! Monsieur Ca-Ce-Ci-*Co-cu.*" — A sarcasm, which was not applied at hazard.

A few more slight anecdotes will mark the manners of that day at Lyons, and the good and bad qualities apparent in the different ranks of society. An English gentleman, who seemed to be very popular among his companions, had brought himself into sudden distress by an unlucky run at play. He was arrested, while he was entertaining several of his countrymen at dinner. Not one of them interfered in his favour; but when he retired from the room, a valet de place, who had lived with him for two years, offered him a purse, containing more than the debt for which he was arrested, telling him, that he had earned that money by the English, and that it could not be better employed, than by saving a gentleman of that country from disgrace. The offer was accepted, and the English gentleman soon afterwards repaid the sum, with the addition of a handsome present.

Another instance of generosity, shewn to an Englishman in distress, occurred while I was at Lyons. A gentleman was arrested for numerous debts, which he had incurred by living in a very extravagant manner with Mad^{elle} St. Clair, an actress of great celebrity and some beauty. She had fascinated the gentleman so completely, that he had lavished upon her all the money, and had exhausted all the credit, which he could command. Tradesmen to whom he was indebted, becoming acquainted with his situation, found it necessary to enforce payment by securing his person. None of the English came forward to his assistance, and he was actually placed in confinement. He was not, however, left long in this situation; for Mademoiselle St. Clair sold all her plate and jewels, and released him. When her lover flew to her, to express his gratitude, he was astonished to find a reception very different from what he expected: after expressing in the fondest manner her affection, she declared it to be her fixed determination, to live with him no longer. In vain he pleaded his constancy, his entire devotion to her wishes. She acknowledged all his claims, but steadily refused to continue a connexion, which must necessarily end in his ruin. She had given such a signal proof of her disinterestedness and affection, that no mercenary motive, or any caprice of sentiment, could be attributed to her conduct; she therefore claimed the merit of the greatest sacrifice in giving him up, to preserve him from himself. All

the Lyons world applauded her generosity :
she was caressed and invited to some of the
best houses in that city. I have dined with
her at Madame de Verpillier's, with a large
society of the best company. Had I not
known that she was an actress, I could not
have discovered her situation by any thing
in her manners or conversation.

CHAPTER XII.

————

ABOUT this time a fatal catastrophe, that befel two lovers, made a great noise at Lyons. A young painter, of considerable eminence, came there, in company with a woman of uncommon beauty, who was his mistress. There was something remarkably attractive in both the man and the woman, and their company was sought for with the utmost enthusiasm by all the young men of that city.

The urbanity, liveliness, and good nature of the young painter, were extolled in every company. Both he and the lady sang and played well on several instruments; and, by a variety of other talents, which they exercised without ostentation, they made what is called in France a great *sensation*. Their mutual fondness kept all pretenders to the lady's favour quite at a distance, while it excited a lively interest among their acquaintance.

There was still however something mysterious in their conduct towards each other, that induced an indefinite kind of suspicion. In the midst of gaiety or mirth, a look, or a sigh, betrayed a secret anxiety. This anxiety gradually increased, notwithstanding the pains which were taken to conceal it. After some months, the stranger and his mistress invited all their acquaintance to a handsome supper, which they gave at taking leave of their friends, before their intended departure from Lyons. When they bade farewell, they shewed great emotion, and hastily withdrew before their friends departed.

There is, near a convent at Lyons, a place which was called the tomb of the two lovers.—On this spot the bodies of the strangers were found the next morning.— They had shot each other with pistols, the triggers of which were so connected by a red riband, as to go off at the same moment. At first no trace of their history, or motive for their conduct, could be discovered : but at length it was ascertained, that the man laboured under some incurable disease, to which the physicians had convinced him he must fall a sacrifice within a given period. His mistress had determined to live no longer than her lover : they had therefore converted whatever they possessed into ready money, which they agreed to spend in the manner most congenial to their tastes; and as soon as their funds should be exhausted, which they had calculated would last to the predicted period when his disease must end his life, they had resolved to destroy themselves. They had projected various means of accomplishing this fatal purpose; poison or drowning had been proposed, and had been rejected, because they could not be certain, that they should both cease to exist at the same moment.

Whoever is acquainted with the state of society in France about this time may form some idea of the interest produced by this extraordinary catastrophe. It was the universal subject of conversation. The young and romantic applauded the heroism of the lovers; the courage and constancy of the woman excited sympathy and enthusiastic applause. In particular, the disciples of Rousseau shed tears of the tenderest compassion over their grave, while the wise and good expressed abhorrence of the selfish, inhuman conduct of a man, who could so far abuse the confidence reposed in him by his mistress, as to persuade her, or to permit her to persuade herself, that true love could justify deliberate murder. That the man might, from the fear of pain or poverty, be hurried to suicide, does not appear out of the course of common cowardice or folly; but to destroy the object of his love, merely to prevent her from living after he ceased to exist, and to prevent her from loving another, when he could no longer feel either love or jealousy, was justly considered as the basest and most depraved egoism.

The idea of constancy after the death of a beloved object is cherished in society, because it is thought a means of preventing inconstancy during life. The woman, who looks forward without restraint to a second or a third husband, may consequently, perhaps, be less attentive to the declining health of him, with whom she still lives;

so far the feeling of posthumous fidelity may be advantageous to society : but this by no means justifies the ferocious sentiment of wishing to sacrifice the object of our affections, lest she should confer happiness upon another.

Scarcely had the interest inspired by this romantic history of the painter and his mistress subsided, when a new *sensation*, of a different sort, was excited at Lyons, by the misconduct, I am sorry to say it, of some of my countrymen. While Captain Jervis* and Captain Goodall of the navy, Mr. Grey†, Lord Winchelsea, Lord Fortescue, Lord Morton, and his amiable family, who were then all at Lyons, did honour to their country : there were others who gave unfavourable specimens of the manners of Englishmen. Among several instances, of which I was unfortunately a witness, I shall mention but one, which will sufficiently mark the contempt shewed for decorum, and for the persons whose civilities my countrymen condescended to accept.

* Afterwards Lord St. Vincent.

† Afterwards Lord Grey,—the late Lord Grey.

It was the custom at Lyons, for the principal families to give a ball once a year to their daughters. As young unmarried women were not usually admitted in every day society, this annual ball was always conducted with the most perfect decorum.

Mothers, who were professed coquets at other times, upon this occasion maintained a matronly reserve, and the utmost respect was shewn by all the young men to the innocent objects of the entertainment. The English, resident at Lyons, were usually invited to this ball, and it was generally understood, that great circumspection was necessary in their behaviour.

Lord * * * * * * * * * * * * * * * * * * was then at Lyons, a *roué*, whose faults were more hurtful to himself than to others; generous, extravagant, dissipated, good natured, and good humoured, he spent his time, and a great deal of the Duke his father's money, in what he thought pleasure. Among others he went to this ball; but he went, unlike others, highly intoxicated. After he had been in the ball-room a short time, he laid himself down upon some chairs, and demonstrated

audibly to the company, that he had fallen
fast asleep. This was considered by the
friends and brothers of the young ladies in
the room as an unpardonable affront to the
company. The young men, many of them
military, collected together, to determine
who should have the lead in turning the
offender out of the room. Mons. de Verpillier, son of the late Commandant of
Lyons, was a friend to the English: he saw
what was going on, and he came to me in
great agitation, to acquaint me with the
danger, into which Lord ＊ ＊ ＊ ＊ ＊ had
brought himself. I immediately expressed great indignation, walked with Mons.
de Verpillier to the place where his drunken
Lordship lay, I pulled him roughly by
the collar, and, when he wakened, I demanded from him immediate satisfaction
for the disgrace which he had brought upon
his countrymen, and the injury which he
had done them, by causing, probably, the
exclusion in future of Englishmen from the
select parties of the best companies in
France. Drunk or sober, his Lordship was
always ready to answer such a summons.
By the assistance of his friends, he was *got*
out of the room : he was not in a situation

to fight that night, but he was in a situation to come to me the next morning, to thank me most heartily for my interference. M. de Verpillier, the best natured man living, easily pacified the young officers, who had been offended : they perceived, that no affront was intended ; and the manner in which I interfered was considered by the French as serviceable to their friends as it was to mine.

I had occasion some time afterwards to draw upon the credit in society, which had been granted me by my French friends in Lyons. It happened, that 1 went with several of my countrymen to the installation or appointment of the major of Lyons. The assembly was held in the great saloon of the Hotel de Ville, which is, or was, one of the finest rooms in Europe. It was furnished with some excellent pictures of the former kings óf France ; some indifferent portraits of Lewis the XVth, and of his family; and several daubs of mayors and aldermen. A speech was made, I believe by a young advocate, which he prefaced by various genuflexions, and bows, and salutations to the pictures : beginning with the portraits of the younger part of the royal family, he addressed each of them by their full titles, pompously pronounced, and with appropriate compliments; but at length, as he turned toward the portrait of majesty, he suddenly started, put his hands over his eyes, as if they were dazzled by the effulgence of the royal countenance, and bowing, or rather prostrating himself in the rostrum, he again timidly looked towards the bright object of his devotion, and again withdrew his presumptuous gaze, till accustomed to the blaze of glory, he was enabled to pay his tribute of adoration to the royal image. This farce appeared to me so ludicrous, that I fear the relaxation of my muscles gave offence to some of the spectators; for when I went down the great stairs, by which I had ascended, I was stopped in a very rude manner by one of the centinels, who told me, that I could not pass that way. I civilly represented to him, that I had come in by that passage, that I was a stranger, and that I did not know any other way out of the Hotel de Ville.

"What, do you grumble?" said the centinel: "I'll teach you better!" and, seizing me by the fleshy part of the inside of my arm, he grasped it with the most brutal violence. I called for his serjeant; the spectators took my part, and the serjeant, though unwilling, was forced to call his officer, who was the major of what I now found was the *garde bourgeoise* of the city. With the *bourgeoisie* I had hitherto had no intercourse. The major, instead of expressing any disapprobation of the soldier's conduct, said with great insolence, " the flesh of *vous autres anglois* is so much softer than that of a Frenchman, that you complain of the slightest touch."

I had the prudence to constrain my anger, and only to say, that he might hereafter be convinced of his mistake. Lord Winchelsea, with whom I had but a slight acquaintance, in the most handsome manner immediately offered to go with me to the *commandant* of Lyons, to demand redress. As the commandant, M. de Bellesize, was my friend, I declined Lord Winchelsea's offer; but I requested, that if I had occasion to lay the affair before our ambassador, he would have the goodness to assist me. I went to M. de Bellesize, who expressed much concern at what had happened: with great kindness he referred me to the colonel of the regiment, assuring me, that he was a man of honour, from whom I should receive full satisfaction; and he hinted, that the commandant should, in such a case, be applied to only in the last resort. I found from the Colonel, M. St. Eloi, the reception which had been promised me; he expressed much indignation, and desired me to meet him in the play-house that evening. I was careful to be there before the play began, and placed myself in the most conspicuous part of the house. Before the overture was finished, I saw Col. St. Eloi and my man appear in the opposite box; and upon a signal to me from the Colonel, I went into the lobby, where there were a number of persons of the highest class in Lyons. M. St. Eloi presented the officer to me, and said, that he was come to ask pardon for the rudeness, with which he had treated me in the morning. As the man advanced towards me, I perceived, that he could scarcely believe that I was the person whom he had offended, as I was now, according to the fashion of the times, in full dress; and

in the morning he had seen me in a frock.
He began to speak in great confusion, but
I requested, that he would spare himself the
trouble, as I considered the circumstance
that had happened as extremely advantageous to me, since it had made me acquainted with M. St. Eloi.—So terminated an affair, that might in the hands of
a haughty or precipitate officer have ended
seriously.

While I was at Lyons, I escaped a dangerous overturn, by having fortunately observed a trivial circumstance, in the manner in which some timber lay on the side of
a road. To the south of the city there is,
or there was, a very narrow road on a hill,
on the banks of the Saone. The banks
were so steep and abrupt, as to form quite
a precipice ; but on this road my coachman, and all those acquainted with the
place, were in the habit of driving with
perfect security. As he drove me up the
hill one morning, I observed some poles,
or balks of wood, lying on the edge of the
road obliquely, encroaching on it but little ; indeed it was so narrow, as to admit
but of little encroachment. The ends of

the timber hung over the edge of the precipice. My coachman, proud of his skill,
passed these poles dexterously as we went
up the hill, as they were inclined from the
precipice in the direction of our ascent;
but as we returned, it was another affair.
It was downhill, and the balks were inclined from the road over the precipice, in
the direction in which we were going. I
perceived that, as we ascended, the poles,
which were long, bent away from the pres-
sure of the wheels; and that of course as
the coachman had kept as close as possible
to the bank on the opposite side, the wheels
must now get between some of the poles,
and if this happened, must, before the horses
could be stopped, necessarily be conducted
over the edge of the bank. I called to the
coachman, who, relying on his skill, and ignorant of the danger, turned a deaf ear to
me.—I bawled, and, I am afraid, I swore;
but all in vain, he reiterated, " *Allez ! allez ! ne craignez rien !*" At last, I drew my
sword, and, passing its glittering blade close
to his cheek, uttered with a tremendous
voice, " *Arrêtez !*" He pulled up at the
very moment when the fore wheel got be-

tween the poles. A yard further, we should
have been inevitably dashed to pieces. Of
the whole party, the coachman was the
most frightened, when he perceived how
narrowly he had escaped.

I must not omit a circumstance, which
will shew, that, notwithstanding all the pains
I had taken to correct my naturally passionate temper, it was not always as much
under my command as it ought to have
been. I happened one morning to be present, when a fencing-master had some dispute with Mr. Day, I think about a pair of
shoe-buckles. The man was a coxcomb,
and impertinent; Mr. Day, not perfectly
understanding the French language, did
not feel the force of his opponent's expressions, but coolly repeated his own state-
ments. The fencing-master forgot himself
yet more; he retorted insolently, and put
his hand on his sword, for in those days
fencing masters in France wore swords: I
darted forward, seized his sword, broke it
in two, kicked him down stairs, and threw
the pieces of the sword after him. For this
violence, instead of being reproved by public opinion in Lyons, I was applauded.
The fencing-master afterwards most hum-

bly begged pardon. Mr. Day made him a present of a new sword, and he was satisfied.

Upon another occasion, I was riding with a lady near Lyons, when a carter did not immediately make way for us. I called to him—he made some answer unfit to be made before a lady : I gave him a stroke with my whip. I saw him feel in his pocket for his knife, and instantly I knocked him down, left him in the ditch, and we rode on. Upon this occasion I thought my temper had not carried me too far, and that I had done right, and no more than was right. I was much surprised when I went into company in the evening, to find that all my French friends looked coolly upon me. My adventure of the morning was, I guessed, in some way or other, the cause of this. I inquired what was the matter, and with some difficulty I was informed, that

" I had *failed* towards the lady with whom I had been riding, and towards myself; that I ought to have left the man dead on the spot."

In consideration of my being an Englishman, and ignorant of what French honour required, I was, however, pardoned, and re-instated in public opinion.

During all the time which I spent at Lyons, my attention to the works in which I was engaged never remitted. A new channel for the Rhone had now been cut for a considerable distance, and the river was nearly intercepted, when an old boatman warned me, that a tremendous flood might be expected in ten days from the mountains of Savoy. I represented this to the company, and proposed to employ more men, and to engage, by increased wages, those who were already at work, to continue every day till it was dark. I took care to have my representations entered in the books of the company, but I could not persuade them to a sudden increase of their expenditure. They had laid out a very large sum of money, the time for paying the stipulated percentage on their subscriptions had not arrived, and they preferred a risk which they did not understand, to advancing money on a demand which they had not expected.

At five or six o'clock one morning, I was awakened by a prodigious noise on the ramparts under my windows. I sprung out of bed, and saw numbers of people rushing towards the Rhone. I foreboded the disaster!—dressed myself, and hastened to the river.

A crowd was collected on the rampart; through this crowd I had to make my way. I was known, for it had been among all ranks of people a favourite amusement, to go to look at the works which I was carrying on. At this moment, I was let to pass through the opening crowd with every mark of civility; and from all parts I was told with an earnestness, which I can never forget, that nobody blamed me, that it was publicly known that I had foretold the disaster, and that it would not have happened, if my advice had been followed.

When I reached the Rhone, I beheld a tremendous sight! all the work of several weeks, carried on daily by nearly a hundred men, had been swept away. Piles, timber, barrows, tools, and large parts of expensive machinery, were all carried down the torrent, and thrown in broken pieces upon the banks. The principal part of the machinery had been erected upon an island opposite the rampart. Here there still remained some valuable timber and engines, which might, probably, be saved by immediate exertion. The old boatman, whom I have mentioned before, was at the water-side; I asked him to row me over to the island, that I might give orders how to preserve what remained belonging to the company. My old friend, the boatman, represented to me with great kindness the imminent danger, to which I should expose myself. " Sir," added he, " the best swimmer in Lyons, unless he were one of the *Rhone-men,* could not save himself if the boat overset, and you cannot swim at all."

" Very true," I replied, " but the boat will not overset; and both my duty and my honour require, that I should run every hazard for those, who have put so much trust in me."

My old boatman took me over safely, and left me on the island; but in returning by himself, the poor fellow's little boat was caught by a wave, and it skimmed to the bottom like a slate or an oyster-shell, that is thrown obliquely into the water. A general exclamation was uttered from the shore! but, in a few minutes, the boatman was seen sitting upon a row of piles in the middle of the river, wringing his long hair with great composure.

I have mentioned this boatman repeatedly as an old man, and such he was to all appearance; his hair was grey, his face wrinkled, his back bent, and all his limbs and features had the appearance of those of a man of sixty, yet his real age was but twenty-seven years. He told me, that he was the oldest boatman on the Rhone; that his younger brothers had been worn out before they were twenty-five years old. Such were the effects of the hardships, to which they were subject from the nature of their employment. They usually lived six or seven hours a day up to their middle in water, with their heads exposed to the burning sun: but their wages were enormous, and supplied them copiously with brandy.

No stimulus seems to have such mighty power upon the uneducated classes of mankind as spirituous liquors. Fear, hope, ambition, shame, avarice, and even mighty love, give place to the pleasures of inebriation! This is a subject worth the pains of accurate investigation.

After my arrival at the island, I had it in my power to preserve a considerable property belonging to the company, which would have been either carried away by the flood, or pillaged by the populace, had I not interfered. At my return to my apartments, I found a note from M. Bono, a banker of Lyons, requesting to see me immediately. I went to his house, where he was just sitting down to his frugal dinner; he asked me to *take soup* with him; and, when we were alone together, he said to me,

" Sir, by the misfortune which has happened to day, you may possibly be put to some immediate inconvenience for a supply of money ; I have a thousand louis d'ors at your service, which you may draw for at sight."

At this time I had good reason to believe, that I was known at Lyons only as a traveller, who had credit for a certain sum on a banker, without any inquiry having been made concerning my property, or my private circumstances. This instance of M. Bono's generosity and confidence made a

great impression on my mind : it was a
practical lesson against that species of national prejudice, which induces men to
think, that virtue is to be found only in
their own country. Whoever is enslaved
by prejudice, like every other slave, has
lost half his manly worth.

I did not accept of M. Bono's generous
offer, because I had no occasion for the
money. My services as an engineer had
been entirely gratuitous, and I had no connexion whatever with the monied concerns
of the company engaged in these public
works ; therefore the disaster which had
happened could not affect me in a pecuniary way. But of this M. Bono was
wholly ignorant; he knew nothing of the
terms on which I stood with the company,
and I might probably, for any thing he
could guess to the contrary, have been by
this accident reduced to the utmost distress. That it happened to be otherwise,
neither diminishes his generosity, nor my
grateful sense of the obligation.

The winter now put a stop to our works
on the Rhone ; but a piece of land had already been gained, and on this site flour-
mills were to be built. On the profit of

these mills the company chiefly depended for reimbursement of the large funds, which they had laid out. During the winter, I employed myself in drawing plans, and trying experiments, preparatory to building flour-mills. But I was not to carry any of these designs into execution.

I mentioned that Mrs. Edgeworth had returned to England, to be confined. In the month of March, I heard that she was brought to bed of a daughter (my daughter Anna). A few days afterwards I received a letter with an account of my wife's death, and I was obliged immediately to return to England.

I had finished an essay on the subject of mills, and as it was ready to be presented to the. company, I sent it to them with a letter, stating the necessity of my departure. In a very handsome manner they sent me a deed conferring upon me a lot of ground, in the new town, which we had won from the ancient conflux of the Rhone and Saone. M. Rigaud de Terrebasse was my trustee: the laws of France at that time preventing foreigners from enjoying this species of property. But the revolution has swept away the family of Terrebasse, and the remembrance of me and my services.

I pursued my journey, taking my little boy with me : we travelled through Burgundy to Paris. The country revived my health, which had suffered much by my assiduous attendance upon the works on the Rhone the preceding summer. I staid a short time at Paris, where M. de Bellesize was then Prévôt des Marchands. He had been removed from Lyons to that important post, upon account of his superior merit. He shewed me great civility, and offered to procure for me the ribbon of the order of St. Michael. This order had been established, I believe, by Lewis the Fourteenth : at first it had been bestowed only on merit. It had degenerated, however, like every thing else under the reigns of the mistresses of Lewis the Fifteenth ; and though it still retained its credit as to foreigners, I declined accepting it.

CHAPTER XIII.

———

A NEW æra in my life was now beginning. When I arrived in London, I found a letter from my steady friend, Mr. Day, promising that he would immediately come from a remote part of England to meet me. I appointed the place of our meeting at Woodstock, where I could breathe the fresh air of Blenheim, while I waited for his arrival.

The first words he said to me were,

" Have you heard any thing of Honora Sneyd?"

I assured him, that I had heard nothing, but what he had told me when he was in France. That she had some disease in her eyes, and that it was feared she would lose her sight; I added, that I was resolved to offer her my hand, even if she had undergone such a dreadful privation.

" My dear friend," said he, "while virtue
and honor forbade you to think of her, I
did every thing in my power to separate
you ; but now that you are both at liberty,
I have used the utmost expedition to reach
you on your arrival in England, that I
might be the first to tell you, that Honora
is in perfect health and beauty ; improved
in person and in mind, and, though surrounded by lovers, still her own mistress."

At this moment I enjoyed the invaluable
reward of my steady adherence to the resolution, which I had formed on leaving
England, never to keep up the slightest
intercourse with her by letter, message, or
inquiry.—I enjoyed also the proof my
friend gave me of his generous affection.
Mr. Day had now come several hundred
miles for the sole purpose of telling me of
the fair prospects before me ; and this proof
of his friendship, and of his being absolutely incapable of envy or jealousy, was
shewn at a moment, when he, from some
trifling cause, thought he had reason to
complain of my having neglected him.

I went directly to Lichfield, to Dr. Darwin's. The Doctor was absent, but his

sister, an elderly maiden lady, who then kept house for him, received me kindly.

"You will excuse me," said the good lady, "for not making tea for you this evening, as I am engaged to the Miss Sneyds; but perhaps you will accompany me, as I am sure you will be welcome."

Pardon me, benevolent reader, if I dwell, for half a page, upon a subject which can but little interest any human creature except myself; we all of us acknowledge, that love "has been, or ought to be, our master."

It was summer—We found the drawing-room at Mr. Sneyd's filled by all my former acquaintances and friends, who had, without concert among themselves, assembled as if to witness the meeting of two persons, whose sentiments could scarcely be known even to the parties themselves.

I have been told, that the last person whom I addressed or saw, when I came into the room, was Honora Sneyd. This I do not remember; but I am perfectly sure, that, when I did see her, she appeared to me most lovely, even more lovely than when we parted. What her sentiments might be, it was impossible to divine.

My addresses were, after some time, permitted and approved; and, with the consent of her father, Miss Honora Sneyd and I were married (1773) by special license, in the ladies' choir, in the Cathedral at Lichfield. Mr. Seward, under whose care (which had been the care of a parent) Honora had been brought up, married us. The good old man shed tears of joy, while he pronounced the nuptial benediction. Mrs. Seward shewed us every possible mark of tenderness and affection; and Miss Seward, notwithstanding some imaginary cause of dissatisfaction which she felt about a bridesmaid, was, I believe, really glad to see Honora united to a man, whom she had often said she thought peculiarly suited to her friend in taste and disposition. Immediately after the marriage ceremony we left Lichfield, and went to Ireland.

Note by the Editor.

I cannot forbear inserting Mr. Day's letter of congratulation on my father's marriage:—

August.

" I am afraid, my dear friend, you have thought that my congratulations were slow on their journey to Ireland; and that I might, long ere this, have sent you an epithalamium in heroics; but excuse my indolence, an indolence I pardon myself, because I had nothing particular to say, and have been in continual motion. I now send you the sincerest wishes, that you may, in this marriage, continue to find every good and comfort you expect; and this is as much as friendship can wish, and more than, according to the common fate of men, will be your share. Be happy, my dear friend, since you possess every thing, which your own mind suggests to you as a means of happiness: more you cannot have. You possess an understanding improved by observation, goodness of temper, and a variety of literary tastes. Whatever my friendship may be able to contribute to your happiness, you know you may command. I am arrived at that period of life, when, in a reflective mind, its sentiments are not easily changed; and if my present aversion to all engagements, which gradually involve the mind in low pursuits, continue, still more may it be presumed, that nothing will ever happen to destroy the strong desire I feel, both from reason and nature, to discharge with propriety all the duties of a man. But you must consider, that, though our affection remain as strong as ever, our habitual intercourse must necessarily be diminished. When you experienced vexations, you sought a comforter in me, and I hope sometimes succeeded: to me you entrusted your uneasiness, your hopes, your fears, your passions. Young and inexperienced in the world, I was capable of being of little use to you, except by fidelity and discretion. To you, when my hopes were more active, and life a novelty, I entrusted all the fantastic emotions of my own heart —schemes of happiness, which a young man conceives with enthusiasm, pursues with ardor, and sees dissipated for ever, as he advances. You heard me with kindness—sometimes repressed, sometimes excited me: in general advised me well, and never deceived my expectations or my confidence. For all which accept the *desire* I have always had to serve you for the *deed*, and receive my thanks—not as a vain acknowledgment, but as a testimony that I am contented with the past.

" You must perceive, that the tendency I have to stoicism, joined with the change of your circumstances, and your acquisition of an amiable friend in a wife, must necessarily make us of less active importance to each other. You two will be settled in one spot, while I am roving about the habitable earth, not in pursuit of happiness, but to avoid ennui. These circumstances will necessarily prevent, at least for some time, the continuance of our intimacy upon the same terms as formerly.

" Helvetius, always systematizing, affirms, that all friendship is founded upon our wants alone; and that, as these cease, or are satisfied, we gradually forget our friends; but if I may judge from my own heart, as well as he from his, there is a certain, stable, dispassionate affection, which may subsist even in years of absence—a tender remembrance of those we have loved, even had we lost the hope of ever seeing them again: an attention to their happiness, were we assured they could never be informed of it. It is therefore in absence, that we form the most accurate judgments of our own hearts; that we separate the local, accidental, temporary affection, from that which is of a stable nature; and this is that affection, which I hope you will retain for me, and which I shall, if I can speak of any future event with certainty, always be sensible of for you. As I know not your present scheme of life, and only know of my own, that I do not at present intend to settle any where, I think it may be some time before we meet again. Whenever we meet, I can receive no greater pleasure, than to see you happy in your wife, your children, and, above all, in yourself.

" As to myself, I can, I think, give no better picture of my own mind, than what I wrote to you last winter from France: an indifference to all human affairs, an aversion to restraint, and engagement, and embarrassment, continue to increase in my mind; so that there is great probability I am marked out by fate for an old bachelor, and an humourist, destined, perhaps, to become very old, because I am very indifferent about the matter, and to buy hobby-horses for your grandchildren ; and, perhaps, as an old friend to the family, admitted to mediate for some of the future Miss Edgeworths, when they run away with a tall ensign in the guards, or their dancing master !

" When I consider your situation in life, I think it probable, with proper attention to your affairs, that, though you have acted precipitately, you have chosen the most eligible manner of living.

" Doctor Small, with whom I now am, has shewn me part of a letter that he has received from you, which gives me real pleasure; nor have I any doubt that the lady, with whom you are now connected, will never give you more uneasiness, should her health continue good, than she has done this first month of your connexion. From her I expect to see how a sensible and affectionate woman should behave to her husband, her husband's children, and her own (no easy task); as in you I hope to find the example of a husband's discretion with a lover's tenderness. With what pleasure shall I, when I meet you again, contemplate that happiness, which you say you so fully possess ! such sights are sometimes necessary to reconcile me to the mass of misery I see around me.

" Pray, to Mrs. Edgeworth say from me every thing, that

may best express the real friendship and esteem I have for
her, and the conviction, that, so far from being any obstacle
to our future friendship, she will always entertain for me
such sentiments, as I deserve from my behaviour to her
husband.

" Adieu.

" T. D."

CHAPTER XIV.

ON our arrival at our home at Edgeworth-Town, we found that much was necessary to be done.

We had a tolerable house, built according to the taste of the foregoing half century, when architecture had not been much studied in Ireland. The grounds and gardens were in a style correspondent to the architecture; the people were in a wretched state of idleness and ignorance: so that we had sufficient business to occupy the whole of our thoughts for some months. My wife applied herself instantly to the various occupations which surrounded her. We had brought with us some English servants, who soon put our domestic economy upon a comfortable footing. The axe and the plough were presently at work. The yew hedges, and skreens of clipped elms and horn-beam, were cut down to let in the air, and the view of green fields. Carpenters and masons pulled down and built up.

As to the people, we perceived, that, by kindness and attention to their wants and prejudices, it would be in our power to meliorate the condition, and to improve the disposition of the poor in our neighbourhood. Few gentry lived near us; fortunately the few, who were within our reach, were friends and relations. But we did not at this time feel the want of society.

A trifling circumstance happened soon after we were married, which the reader will smile at my recording. But " If trifles light as air are to the jealous confirmation strong," they have also as powerful an effect as proofs of affection. Mrs. Edgeworth one day missed her wedding ring. The emotion which she shewed at this slight accident, an emotion so foreign to her usual habits of mind, and the eagerness, and anxiety, yet the presence of mind, with which she instantly pursued the best means to find this ring, were proofs of attachment, which I might scarcely be pardoned for mentioning at this period of my life, except as they are connected with a circumstance, that happened several years afterwards.

The moment she perceived, that the ring was not on her finger, she started up from dinner, and went to the place where she had been standing during most part of the morning, and where thirty or forty laborers were employed levelling a piece of ground. She desired them to stop, and to range themselves in one line. She then, promising them a considerable reward if they found her ring, desired them to walk regularly one after another, as slowly as possible, across the field, without separating, lest they should tread the ring into the fresh ground. She obliged them to sift all the newly turned earth, stood by, and examined each sieve-full herself. The ring was found, and Honora vowed, that *she never would lose it again but with her life.*

During the first years of our marriage we were much alone; and as we were both fond of a country life, and of each other, we felt too well contented in retirement. I say this, because it is a fault in which young people, who have loved one another with strong passion, are apt to indulge; they are inclined to seclusion, and think that two real lovers are all-sufficient to each other. That we did enjoy great and *untired* felicity is certain; but at the same time I acknowledge, that for three or four years we did not improve our understandings, or enlarge our views of life, as we should have done, had we lived in a more extended society. In another point of view, it may be observed, that the retired domestic manner, in which we passed our time, kept dormant the seeds of ambition, which might probably have sprung up in my mind, if I had frequented the capital, where I had many contemporaries eager in the race of parliamentary distinction, to whom I felt that I was not absolutely inferior.

After having resided three years in Ireland, we determined to pay a visit to our friends in England. We certainly found a considerable change for the better, as to comfort, convenience, and conversation, among our English acquaintance. So much so, that we were induced to remain in England.

We took a house in Hertfordshire, at North Church, near Great Berkhampstead. The house was small, but uncommonly neat and cheerful. Here we resolved to pass two or three years in retirement; to cultivate our tastes for literature, and to give me leisure to pursue my propensity for mechanics. As we were within five and twenty miles of town, we had hopes, that I should be able to renew and sustain my acquaintance with those men of science and literature, with whom I had formerly been intimate.

Mr. Day continued to keep up a strict intimacy with my family; he paid us frequent visits in our retirement. He had purchased chambers in the Temple, where he spent a considerable portion of his time, and whence he made excursions to different parts of England.

His attachment to Miss Elizabeth Sneyd had long since terminated unsuccessfully. Upon his return from France, that lady found, that, notwithstanding all the exertions he had made, and the pains he had taken to improve his manners and person, she could not feel for him the sort of attachment, which was necessary for her happiness and for his in marriage. In short, notwithstanding his great and good qualities, she could not give him her heart. This disappointment affected him much at the moment; but time produced its usual effect. In the course of four or five years his heart had been again at liberty—and again engaged. His fate in marriage was at last decided by his friend, Dr. Small.

I have said, that Mr. Day was the man of the most perfect morality, whom I have ever known. Dr. Small was in this respect scarcely his inferior, and he was far his superior in knowledge of the world, in experience, true philosophy, and suavity of manners. He had acquired over most of those who were intimate with him an indescribable ascendance: over the mind of Mr. Day his influence was paramount to that of any other individual, and it was used solely for his friend's advantage. As marriage was the grand object of Mr. Day's consideration, Dr. Small was constantly interested upon this subject. He never

saw any woman, whose character and situation in life appeared suitable to Mr. Day,
without mentioning her to him, and endeavouring to give him an opportunity of
judging for himself. The Doctor accidentally became acquainted with Miss
Milnes, of Wakefield, in Yorkshire; and,
soon after he had seen her, he told me, that
he believed he had at last found a lady perfectly suited to Mr. Day; a woman, who
was capable of appreciating his merit, and
of treating the small defects in his appearance and manners as trifles beneath her
serious consideration.

But the Doctor did not hastily communicate his sentiments to his friend, for Mr.
Day had by this time become attached
to Sabrina. She had now grown up, and,
no longer a child, was entitled by her
manners and appearance to the appella-
tion of a young lady. Mr. Day took great
pains to cultivate her understanding, and
still more to mould her mind and disposition to his own views and pursuits. His
letters to me at this period were full of
little anecdotes of her progress, temper,
and conduct: I had not formerly thought,

that she was sufficiently cultivated, or of
a sufficiently vigorous understanding, to
be his companion. I knew also, that whoever should become the wife of Mr. Day
must be content to live in perfect retirement; to give up her tastes to his; to
discuss every subject of every day's occurrence with logical accuracy; to be totally indifferent to all the luxuries, and to
some of the comforts of opulent life. To
balance these sacrifices, she would find herself united to a man of undeviating morality, sound sense, much knowledge, and
much celebrity; a companion never deficient in agreeable or instructive conversation, of unbounded generosity, of great
good-nature; a philanthropist in the most
extensive, and the most exalted sense of
the word: in short, a man who would put
it in her power to do good to every body
beneath her, provided she could scorn the
silly fashions of those above her. Sabrina
was, as to many of these circumstances,
well suited to Mr. Day; but she was too
young and too artless, to feel the extent of
that importance, which my friend annexed
to trifling concessions or resistance to fashion, particularly with respect to female

dress. He certainly was never more loved
by any woman, than he was by Sabrina;
and I do not think, that he was insensible
to the preference, with which she treated
him; nor do I believe, that any woman
was to him ever personally more agreeable.

From his letters at this time I was persuaded, that he would marry her immediately;
but a very trifling circumstance changed
his intention. He had left Sabrina at the
house of a friend under strict injunctions
as to some peculiar fancies of his own; in
particular, some restrictions as to her dress.
She neglected, forgot, or undervalued something, which was not, I believe, clearly de-
fined. She did, or she did not, wear certain long sleeves, and some handkerchief,
which had been the subject of his dislike,
or of his liking; and he, considering this
circumstance as a criterion of her attachment, and as a proof of her want of strength
of mind, quitted her for ever! The circumstances of this singular transaction and determination I learned from the gentleman,
at whose house they happened. Mr. Day,
at the moment, wrote me a letter explain-
ing to me the feelings and reasoning, which

decided him to give up, from a motive apparently so trifling, a scheme upon which he had bestowed so much time and labour; a scheme which he had recurred to after every disappointment; and which, at last, from the surprising improvement that hope had wrought in Sabrina's mind and manners, promised him a companion, peculiarly pleasing to him in her person, devoted to him by gratitude and habit, and, I believe, by affection. Mr. Day's reasons for breaking off this attachment proved to my understanding, that, with his peculiarities, he judged well for his own happiness; but I felt, that, in the same situation, I could not have acted as he had done.

When Mr. Day's friends were convinced, that he would never again resume his connexion with Sabrina, Doctor Small by degrees opened his views relative to Miss Milnes. The unbounded charity and benevolence of this lady were so well known in Yorkshire, and were so much talked of, that it needed but little inquiry to be certain of the facts. Several of her letters had been seen, which evinced the superiority of her understanding. This was

so generally admitted among her acquaintance, that to distinguish her from another
Miss Milnes, a relation of hers, celebrated
for beauty, who had, I believe, been called
Venus, she had acquired the name of Mi-
nerva. All this Doctor Small reported to
Mr. Day.

" But has she white and large arms ?" said
Mr. Day.

" She has," replied Dr. Small.

" Does she wear long petticoats ?"

" Uncommonly long."

" I hope she is tall, and strong, and
healthy."

" Remarkably little, and not robust.—
My good friend," added Dr. Small, speaking in his leisurely manner, " Can you possibly expect, that a woman of charming
temper, benevolent mind, and cultivated
understanding, with a distinguished character, with views of life congenial with your
own, with an agreeable person and a large
fortune, should be formed exactly according to a picture that exists in your imagination ? This lady is two or three and
twenty, has had twenty admirers ; some of
them admirers of herself, some, perhaps, of

her fortune; yet, in spite of all these admirers and lovers, she is disengaged. If
you are not satisfied, determine at once
never to marry."

" My dear Doctor," replied Mr. Day,
" the only serious objection, which I have
to Miss Milnes, is her large fortune. It
was always my wish, to give to any woman
whom I married the most unequivocal
proof of my attachment to herself, by despising her fortune."

" Well, my friend," said the Doctor,
" what prevents you from despising the
fortune, and taking the lady ?"

Mr. Day soon went into Yorkshire, was
charmed with Miss Milnes, and began a
courtship, which any other man would
have concluded in a few months ; for the
lady not only approved, but admired the
noble and disinterested character of her
lover, and he was equally pleased with her
generous and romantic disposition. When
I say *romantic*, I do not mean that species
of romance, which fills the pages of a novel;
but that romantic forgetfulness of *self*, which
is seldom found in real life, and which had
appeared in every part of the conduct of

Miss Milnes from her childhood. With
Mr. Day there were a thousand small pre-
liminaries to be adjusted : not content with
that influence, which his merit and his superior understanding must necessarily obtain, there was no subject of opinion or
speculation, which he did not, previously
to his marriage, discuss with his intended
bride. She was not only a woman of abilities, but she was conversant with literature, and particularly with poetry, which
she read with much energy ; besides, she
was really mistress of the English language,
and she spoke with great eloquence. Between a lady so accomplished, and such a
dialectician as Mr. Day, it was not likely,
that conversation should languish. In
fact, I believe, that few lovers ever conversed or corresponded more, than did my
friend and Miss Milnes. At length they
were married, and their fortunes were mutually settled upon the survivor in case they
had no children.

To prevent all temptation of being drawn
into the usual modes of life followed by
people in his situation, they retired to a
small lodging at Hampstead, where he de-

termined to reside, till he should find a house suited to his taste.

Shortly after their marriage, he brought Mrs. Day to Northchurch to see us. Her person and conversation were pleasing, and the noble and generous sentiments which she expressed, and the conformity of all her conduct to these sentiments, entitled her to more than common admiration and respect. Mrs. Edgeworth had been well accustomed to Mr. Day's habits of discussion and declamation : she observed that Mrs. Day's replies, replete with sense and spirit, were always delivered in chosen language, and with appropriate emphasis. My friend proceeded towards his conclusions with unerring logic, and inflexible perseverance ; but Mrs. Day's eloquence won the hearers, at least for a time, to her opinions. Mrs. Honora Edgeworth's conversation was in a medium between both styles; she never harangued, but she spoke with ease, and yet with much precision, so that in summing up the arguments on both sides she generally gave a sound judgment.

Notwithstanding the dryness of political

and metaphysical subjects, which were usually those upon which we descanted, I was amused and instructed, and I wished most heartily to prevail upon Mr. Day to settle in my neighbourhood in Hertford-shire; but he had an insurmountable objection to any situation near his former friends, lest, as I supposed, any opinions contrary to his system of connubial happiness might be supported before his wife. He remained some time at Hampstead, being in no great haste to purchase a house; as he thought, that by living in inconvenient lodgings, where he was not known, and consequently not visited by any body except his chosen few, he should accustom his bride to those modes of life, which he conceived to be essential to his happiness.

My wife and I went to see the new married couple, at Hampstead. It was the depth of winter; the ground was covered with snow, and to our great surprise, we found Mrs. Day walking with her husband on the heath, wrapped up in a frieze cloak, and her feet well fortified with thick shoes. We had always heard that Mrs. Day was particularly delicate; but now she gloried in rude health, or rather was proud of having followed her husband's advice about her health; advice, which was in this respect undoubtedly excellent.

I never saw any woman so entirely intent upon accommodating herself to the sentiments, and wishes, and will of a husband. Notwithstanding this disposition, there still was a never-failing flow of discussion between them. From the deepest political investigation, to the most frivolous circumstance of daily life, Mr. Day found something to descant upon; and Mrs. Day was nothing loth to support upon every subject an opinion of her own: thus combining, in an unusual manner, independence of sentiment, and the most complete matrimonial obedience. In all this there may be something, at which even a friend might smile; but in the whole of their conduct there was nothing, which the most malignant enemy could condemn.

Mr. Day afterwards bought a house, and a small estate, called Stapleford-Abbot, near Abridge, in Essex. The house was indifferent, and the land worse; the one he proposed to enlarge, the other to improve, according to the best and latest systems of agriculture. The house was of brick, with but one good room, and it was but ill adapted in other respects to the residence of a family. He built, at a considerable expense, convenient offices; also a small addition to the house.

When Mr. Day determined to dip his unsullied hands in mortar, he bought at a stall "*Ware's Architecture;*" this he read with persevering assiduity for three or four weeks, before he began his operations. He had not however followed this new occupation a week, before he became tired of it, as it completely deranged his habits of discussion with Mrs. Day in their daily walks in the fields, or prevented their close application to books when in the house. Masons calling for supplies of various sorts, which had not been suggested in the great body of architecture, that he had procured with so much care, annoyed the young builder exceedingly. Sills, lintels, door and window cases, were wanting before they had been thought of; and the carpenter, to whose presence he had looked forward, but at a distant period, was now summoned and hastily set to work, to keep the masons a going. Mr. Day was deep in a treatise, written by some French agriculturist, to prove, that any soil may be rendered fertile by sufficient ploughing, when the masons desired to know, where he would have the window of the new room on the first floor. I was present at the question, and offered to assist my friend—No—he sat immoveable in his chair, and gravely demanded of the mason, whether the wall might not be built first, and a place for the window cut out afterwards. The mason stared at Mr. Day with an expression of the most unfeigned surprise. " Why, Sir, to be sure, it is very possible; but, I believe, Sir, it is more common to put in the window cases while the house is building, and not afterwards."

Mr. Day, however, with great coolness, ordered the wall to be built without any opening for windows, which was done accordingly; and the addition, which was made to the house, was actually finished, leaving the room, which was intended for a dressing-room for Mrs. Day, without any window whatsoever. When it was sufficiently dry, the room was papered, and for some time candles were lighted in it whenever it was used. So it remained for two or three years; afterward Mrs. Day used it as a lumber room, and at last the house was sold without any window having been opened in this apartment.

This strange neglect arose from two causes, from Mr. Day's bodily indolence and mental activity; he did not like to get up from his chair to give orders upon a subject, on which he was but little interested, and he felt strongly intent upon the speculation, which then occupied his mind.

Some years after his marriage he bought another house and estate, Anningsley, near Chertsey, in Surry; to which he removed. He told me, that he had well considered the purchase, and that he thought he had acted prudently in buying one of the most unprofitable farms in England, because he had a large scope of ground for a small sum of money. Here he tried, upon a great scale, several of those doubtful processes, which he had found in foreign and domestic books of agriculture, to the no small injury of his fortune: and here he finally settled, at a distance from us, so that we saw him only at long intervals. Whenever he visited us, however, we observed, that he seemed to relax from his usual dialectics, and to let things and persons pass before him unanalysed, and without dissection.

Having thus endeavoured to sketch Mr. Day's character, I feel almost as if I had done an injury, as if I had betrayed some confidence, and added another to the many instances of those, who, by the opportunities of friendship, had acquired the means of injuring a friend. But, upon reflexion, I am acquitted in my own judgment from any unkindness towards him or his. He had been long dead when this was written ; he left behind him no children, nor any near connexions; all our common friends outlived him but a short time, and when this comes before the public, I shall myself be no more. In relating these circumstances, it could not be my intention to throw any ridicule upon my friend, or upon the excellent qualities, which he displayed in every situation, and at every period of his life; but simply to draw a character, which might be useful to others, to shew that *ne quid nimis* (too much of one thing

Thomas Day Esq[r]

Engraved by Henry Meyer
from a Picture by Wright of Derby.

Published March 30th 1820 by R. Hunter, N[o] 72. S[t] Pauls Church Yard. London.

is good for nothing) is a judicious adage—
that there may be too strong an adherence
even to reason; and that discussion upon
all the minute affairs of life, and of every
day occurrence, may become as inconvenient and prejudicial, as the contrary extreme of careless and hasty decision.

I was not disappointed in my hopes, that
during a residence at Northchurch I
should renew my acquaintance and habits
of intimacy with some of my former literary and scientific friends in London. My
mind was kept up to the current of speculation and discovery in the world of science,
and continual hints for reflection and invention were suggested to me.

By some circumstance, which I cannot
now remember, my attention was about
this period turned to clockwork, and I invented several pieces of mechanism for measuring time. These, with the assistance of a
good workman in the neighbourhood, I executed successfully. I then (in 1776) finished a clock * on a new construction. Its accuracy was tried at the observatory at Ox-

* A drawing and description of this clock will hereafter
be published.—*Note by Editor.*

ford, by my friend Dr. Hornsby, who made me a favourable report of its performance. It has been ever since, and is now (in 1809) going well at my house in Ireland.

My wife had an eager desire for knowledge of all sorts, and, perhaps to please me, became an excellent theoretick mechanic. Mechanical amusements occupied my mornings, and I dedicated my evenings to the best books upon various subjects. I strenuously endeavoured to improve my own understanding, and to communicate whatever I knew to my wife. Indeed while we read and conversed together, during the long winter evenings, the clearness of her judgment assisted me in every pursuit of literature, in which I was engaged; as her understanding had arrived at maturity, before she had acquired any strong prejudices on historical subjects, she derived uncommon advantage from books.

We had frequent visitors from Town; and as our acquaintance were people of literature and science, conversation with them exercised and arranged her thoughts, upon whatever subject they were employed. Nor did we neglect the education of

our children : Honora had under her care,
at this time, two children of her own, and
three of mine by my former marriage.

Unfortunately for my eldest son, I was
persuaded by my friends to send him away
from me to school, without having sufficiently prepared him for the change between the Rousseau system, which had
been pursued at home, and the course of
education to which he was to be subject at
a public seminary. His strength, agility,
good humour, and enterprise, made him a
great favourite with his school-fellows ; he
shewed abilities, and was sure to succeed
whenever he applied ; but his application
was not regular, nor was his mind turned
to scholarship. He had acquired a vague
notion of the happiness of a seafaring life,
and I found it better to comply with his
wishes, than to strive against the stream.
He went to sea, readily acquired the knowledge requisite for his situation, and his
hardihood and fearlessness of danger appeared to fit him for a sailor's life*.

* He some years afterwards went to America, married
Elizabeth Knight, an American lady, and settled in South

For some years we continued to pass our time at Northchurch, in the manner I have described, rationally, and therefore happily.

No events of any importance happened. Some slight incidents, however, which occurred to me in my walks through London, remain, after years have elapsed, fresh in my recollection; and yet they are totally unconnected with my own interests or history.

One day, in one of the crowded streets, I met a poor young girl, who seemed utterly bewildered; she stopped me, to ask if I would tell her the name of the street she was in. Her accent was broad Scotch, and her look and air of perfect simplicity was, I perceived, not assumed, but genuine. I gave her the information she wanted, and asked her where she lived, and if she was in search of any friend's house. She said she did not live any where in London; she was but just arrived from Scotland, and knew nobody who had any house or lodg-

Carolina, near George Town. He died (August 1796) leaving three sons, who, with their mother, are still resident in America.—*Note by the Editor.*

ing of their own in Town, but she was looking for a friend of the name of Peggy; and Peggy was a Scotch girl, who was born within a mile of the place where she lived in Scotland. Peggy was in service in London, and had written her direction to some house in this street, but the number of the house, and the names of the master or mistress, had been forgotten. The poor girl was determined, she said, to try every house, for she had come all the way from Scotland to see *Peggy,* and she had no other dependance!

It seemed a hopeless case. I was so much struck with her simplicity and forlorn condition, that I could not leave her in this perplexity, an utter stranger as she evidently was to the dangers of London. I went with her, though I own without the slightest hope of her succeeding in the object of her search; knocked at every door, and made inquiries at every house. When we came near the end of the street, she was in despair, and cried bitterly; but as one of the last doors opened, and as a footman was surlily beginning to answer my questions, she darted past him, exclaiming,

" There's Peggy !"—She flew along the passage to a servant girl, whose head had just appeared as she was coming up stairs. I never heard or saw stronger expressions of joy and affection than at this meeting: and I scarcely ever, for any service I have been able in the course of my life to do for my fellow-creatures, received such grateful thanks, as I did from this poor Scotch lassy and her Peggy for the little assistance I afforded her.

Another time, about this period, one evening in summer, I happened to be in one of those streets that lead from the Strand toward the river. It was a street to which there was no outlet, and consequently free from passengers. A Savoyard was grinding his disregarded organ; a dark shade fell obliquely across the street, and there was a melancholy produced by the surrounding circumstances, that excited my attention. A female beggar suddenly rose from the steps of one of the doors, and began to dance ludicrously to the tune, which the Savoyard was playing. I gave the man some money, and I observed, that, for such an old woman, the mendicant danced with great sprightliness. She looked at me stedfastly, and, sighing, added, that she could once dance well. She desired the Savoyard to play a minuet, the steps of which she began to dance with uncommon grace and dignity. I spoke to her in French, in which language she replied fluently, and in a good accent; her language, and a knowledge of persons in high life, and of books, which she shewed in the course of a few minutes' conversation, convinced me that she must have had a liberal education, and that she had been amongst the higher classes of society. Upon inquiry, she told me, that she was of a noble family, whose name she would not injure by telling her own : that she had early disgraced herself, and that, falling from bad to worse, she had sunk to her present miserable condition. I asked her why she did not endeavour to get into some of those asylums, which the humanity of the English nation has provided for want and wretchedness; she replied, with a countenance of resolute despair.

" You can do nothing more for me than

to give me half a crown:—it will make me drunk and pay for my bed!"

I will not detain the reader with more of my recollections of such anecdotes, as have probably occurred to most men, who have walked much in the streets of London.

CHAPTER XV.

———

In the midst of the domestic happiness
which we enjoyed at this period, I was called away to attend a lawsuit in Ireland.
It was a plain and simple case; but it had
been so managed by my lawyers, that it
became sufficiently complicated, and of
course it was vexatiously protracted. My
wife could not, with tolerable convenience,
accompany me. This was the first time
Honora and I had been separated since our
marriage; a separation extremely painful
to us both!

My lawsuit prospered from the moment
I landed in Ireland; and in less than two
months I had obtained a verdict at the expense of one half the value of the fee-simple of my land. I found, that my estate
had not been improved during my absence;
and that Ireland, and its inhabitants, appeared to less advantage to me, than at any

other time when I had been there. At this
period, from various causes, the rent of land
had fallen, and there was less excitement
to industry than formerly. I thought it
necessary to remain at Edgeworth-Town,
to improve my estate, and thus to sacrifice
my tastes to my duty. I had always
thought, that, if it were in the power of any
man to serve the country which gave him
bread, he ought to sacrifice every inferior
consideration, and to reside where he could
be most useful. I expressed these sentiments in my letters to Honora; she approved of my plan, and she immediately
took measures for letting our house at
Northchurch, and for coming over to join
me in Ireland.

Note by the Editor.

When Mrs. Honora Edgeworth was preparing to leave
England, she received a letter from Mr. Day, which ought
to be preserved as a proof, that he, who on some occasions
disdained the forms of polished life, could, when prompted
by generous friendship, and the real tenderness of his nature,
write with perfect politeness.

"My Dear Madam,

"I have no message to send to Mr. Edgeworth, except
my most affectionate remembrance, and my best hopes,

But, alas! In the midst of her preparation for this journey, Honora's health began to fail. She always attributed the beginning of her illness to a cold she caught one day, when, after a walk to the post-office, she sat down under a tree to rest herself, when she was fatigued and too

---

that whatever he may decide upon may prove as much for his happiness, as I am sure it will for his honour. I am too sensible of your mutual affection, to wish you longer divided; though I should rejoice if the meeting could be in England, which would, I think, be upon many accounts more eligible. However, you are best able to decide for yourself. I will only detain you, to beg that, if you go, you will, upon the account of all your friends, (and more than all put together, of Mr. Edgeworth,) favour the delicate health which you have at present. The greater part of ladies would little need this caution. But it is no compliment from me to say, that you are one of the few women, who, in the discharge of any duty, are rather liable to attempt too much, than to venture too little.

" May you arrive safe in Ireland, and long remain the joy and pride of him, whose lively affection and tenderness, joined to so many other virtues, render him most deserving of so great a good!

" I am, with all respect, esteem, and friendship,

" Dear Madam,

" Your friend and humble servant,

" London,   " THOMAS DAY.
Wednesday Night."

warm. She was seized with a slight fever, and confined to her bed. Her letters told me the exact truth. Truth is, in such circumstances, better than all the arts of concealment and palliation. The physician, who attended her, was not apprehensive of any danger; and as she had in general enjoyed good health during our marriage, I flattered myself he was right. However, I determined to go to her, and immediately went to Dublin; but, just as I was going to embark in the packet, the wind became directly adverse.

Every post brought me letters promising her speedy recovery; yet the anxiety and impatience, which I experienced, were beyond all bounds. I sat, hour after hour, with my eyes fixed upon a weathercock, watching its slightest motions. Day after day the wind continued in the same point; no vessel of any sort could sail. It was moonlight, and whenever I wakened, I got up to look whether the wind had changed. For more than a week I was thus kept in suspense. I almost thought, that I could walk upon the sea. If a row-boat could have been procured, I should not have hesitated to attempt a passage. I had felt the violence of love before marriage, but I assure my reader, that I felt it more strongly after I had been married six years. And, looking back to my past life, I can assert, that the greatest torment my mind ever endured was at this period; and the most ardent and exquisite pleasure I ever enjoyed was on meeting my wife on the road, near Daventry. I found her weak, flushed, feverish, but still the same Honora; the same prudence, the same strength of mind appeared in whatever she said. In a few minutes we determined, as my house was let for a year, to proceed to Lichfield to Mr. Sneyd's, (my father in law,) till we could decide on our future plans.

Here I consulted my friend, Doctor Darwin, from whose manner I soon perceived, that he thought Honora's illness more serious, than I had in the least suspected. I found, that he thought she had a tendency to consumption, a disease from which she had escaped at fifteen. He advised me not to take her to Ireland, but to find some place near Lichfield, where she might have his friendly assistance. Mr.

Sneyd had engaged himself from home
during the summer, and he kindly lent me
his house during his absence. Doctor
Darwin shewed the most earnest attention
to his patient. At times he thought the
disease, which came on slowly, might be
averted before it had taken root in her constitution, and at others he spoke with so
much reserve, as to alarm her most sanguine friends. Her disease encreased
slowly, with all those vicissitudes, which
excite alternate hope and fear, and which
keep the mind in the most painful anxiety.

I had the utmost reliance on the skill and
attention of Doctor Darwin. His enemies,
for merit must excite envy, always hinted,
that he was inclined to try experiments
upon such patients, as were disposed to
any chronic disease. I had frequent opportunity of knowing this to be false; and,
in the treatment of Mrs. Edgeworth, he
never, without the entire concurrence of
her friends, followed any suggestion, even
of his own comprehensive and sagacious
mind, that was out of the usual line of
practice; on the contrary, it was always
in the most cautious, I may almost say

the most timid manner, that he proposed any thing, which he thought beyond the established limits.

Every physician of eminence in England was consulted, either in person or by letter. I took her to Mr. Day's house, where, from its vicinity to London, she might have access to the best advice. The moment Doctor Heberden saw her, I perceived by his looks, that he had no hopes; and when he gave his opinion, he did not attempt to deceive me. He was a man of much sagacity and experience, and his manner of communicating to me that dreadful sentence, which he was obliged to pronounce, was humane, kind, and impressive. He told me, that it was too frequently his lot, to see the most amiable and beautiful, and the most highly endowed females, the most exposed to this incurable malady. He advised me to attend more to the ease of my beloved wife, and to the palliation of symptoms, than to the pretensions of any medical help, that held out hopes of recovery. He said, that, in conversing with us, he saw the strength of our attachment, and the superiority and

uncommon powers of her mind: he observed, that on these I should rely for support, both for her and for myself.

We passed a few days in Essex with Mr.
and Mrs. Day. They were very kind to
Mrs. Edgeworth; but they did not see
her danger in as strong a light as I did.
On our return to Lichfield, where we staid
for some time, the affection of her family,
the conversation of some of her former
friends, and the attentions of all her acquaintance, were sources of great satisfaction to us both. After some months, when
Mr. Sneyd returned home, we took a small
house at Bighterton, near Shiffnal, in Shrop-
shire. Here we had the medical assistance
of Mr. Young, a humane and skilful surgeon: we had also easy access to Doctor
Darwin, and to Mr. Sneyd's family.

Having parted with my house at Northchurch, I found myself happy in being
able to procure any immediate residence,
until it should be determined whether Honora should be advised to go to Clifton,
Devonshire, or some warmer climate. She
seemed willing to try Clifton, which, at
that time, was in high repute; but the

thought of going to any distant country was extremely painful to her.

Bighterton was near Weston, the seat of Sir Henry Bridgman*. Lady Bridgman and her daughters were as kind to Honora as it was possible. Every thing, that their magnificent and hospitable mansion could afford, was at all times at our command. I cannot record this instance of benevolence, without observing that both of us, particularly my poor wife, felt interested by the amiable manners and cultivated understandings of this family. Nor was theirs a sudden burst of kindness, prompted by compassion and sympathy—it lasted— for it arose from generous hearts and sound judgment, that knew how to feel, and how to appreciate the pleasure of doing good.

Honora's disease advanced imperceptibly; but still it advanced. Her sister, Charlotte Sneyd, attended her with unceasing, but with unobtrusive care. Two of my children were with her, on whom their mother continued to bestow the most

* Afterwards Lord Bradford, father to the present Earl.

judicious attention. Her understanding, indeed, never partook of her disease; it was always calm, and clear, and active.

At length, in the month of April, she drew near her end. She had long known the nature of her distemper, and that it was incurable; but never for one moment did pain, or fear, or vain regret for every enjoyment that a human creature could possess, disturb the firmness of her mind. She not only looked upon death, as every wise and good person should behold it; but its near approach did not appal her.

Three days before she died, I was suddenly called up to her room. I found her in violent convulsions. Youth, beauty, grace, charms of person, and accomplishments of mind, reduced to the extreme of human misery, must have wrung the most obdurate heart. What must her husband feel at such a moment?—I felt her pulse, and whispered, " You are not dying." She looked at me with an effort of resolution and kindness, to thank me. When the fit ceased, she begged of me to sit down beside her bed. I took out my pencil, and determined to note whatever she said and

did at this awful period, an employment
that might enable me to bear with more
fortitude the scene that I was to witness.
She soon fell asleep, and wakened smiling.
" I am smiling," said she, " at my asking
you to sit beside me as a sort of protection,
and at my being afraid to die in my sleep,
when I never felt afraid of dying when
awake." The ensuing days she talked,
during the intervals of dosing, about the
education of her children, and about every
thing which concerned my happiness. She
recommended it to me in the strongest
manner, to marry her sister Elizabeth.

After my having sat up all the night
of the 30th of April, I was suddenly
called at six o'clock in the morning.
Her sister Charlotte was with her. The
moment that I opened the door, her eyes,
which had been fixed in death, acquired
sufficient power to turn themselves towards
me with an expression of the utmost tenderness. She was supported on pillows.
Her left arm hung over her sister's neck
beyond the bed. She smiled, and breathed
her last!

At this moment I heard something fall

on the floor. It was her wedding ring, which she had held on her wasted finger to the last instant—remembering, with fond superstition, the vow she had made, never again to lose that ring but with life. She never moved again, nor did she seem to suffer any struggle.

Thus died Honora Edgeworth. The most beloved as a wife, a sister, and a friend, of any person I have ever known. Each of her own family, unanimously, almost naturally, preferred her. My sister, Mrs. Ruxton, loved her with enthusiastic attachment. All her friends adored her, if treating her with uniform deference and veneration may be called adoring.

A dreadful duty still remained to be performed. She had requested, that no person but her sister Charlotte, myself, and her servant, should be permitted to touch her corpse. She was obeyed. That excellent sister, one of the most tender and gentle creatures upon earth, obeyed!—and through all this trying duty shewed a strength of mind and perseverance, that could scarcely be expected from the most resolute of another sex.

My steady friend, the steady friend of
my lost wife, my sister Ruxton, had set
out from Ireland with her husband, who
had the kindness to attend her. But she
arrived, alas! a day too late.

I passed the night before the funeral beside her coffin. The last look!—How often
I thought, that I had taken the last look!

What a variety of recollections and
ideas passed through my mind during this
night! There was a clock in the room. I
observed, that it was four o'clock. I was
standing on the hearth, not leaning against
any thing, and in that position I suddenly
fell fast asleep, and though I slept but five
minutes it seemed to me hours. The day
began to dawn. I opened the window
shutters and the sash; "The lark sang out,
the morning smiled;" a strange sense of life
and renewed vigour came over all my
senses.

Providence has ordained the vicissitudes
of day and night to act upon the human
frame with unceasing influence. In the ordinary course of life, this influence is scarcely
noticed; but upon extraordinary occasions

we perceive its force, without being in the least able to account for it.

I roused myself to bear with fortitude the interment of my wife. The Bridgman family obtained permission to have her buried in the porch of the church, at King's Weston. I followed the coffin. I saw several persons in the church, whom melancholy curiosity had drawn together at an early hour. I heard the service. I saw the remains of what I had so much loved committed to the grave; but when I heard a shovelful of earth thrown on the coffin, a new sensation of grief struck me with unexpected force. I did not shed a tear, or breathe a sigh; but I felt, that I must not stay. I left the church, and found myself at home, I could not well tell how.

My sister Ruxton, and Charlotte Sneyd, and my children, recalled me to myself; and, without lamentation, we talked by intervals of the views before me—of my friends—my children—with a vacancy of mind, that I cannot by any means account for or describe.

After a few days, I went to Lichfield,

where I remained some time. In consequence of a most affectionate letter from Mr. Day, I took my two youngest daughters, Anna and Honora, with me, to his house in Essex. Mr. Day's conversation always interested me. Sound sense and rational conversation solace the mind more effectually, than the most affectionate condolence or sympathy. But even divine philosophy was in vain; I listened, I replied—I endeavoured to read—I felt Mr. Day's kindness to my children—but nothing relieved me from that listlessness, which succeeds to repressed sorrow. From some false notions of manly firmness, I had repressed the natural expressions and consequences of sorrow. The prodigious force, which I exerted over my mind, was, as usual, succeeded by languor. I merely existed, and I felt indifferent about every thing, and every place; though I imagined from theory, that change of place would relieve me.

I returned to Lichfield, where the associations of former times awakened feelings of tender sorrow, which, however sad, were far preferable to that inanimate state, in which I had passed the preceding months.

Note by the Editor.

Among my father's papers, I found in his hand-writing a character of Mrs. Honora Edgeworth, ending with these words:

" The following letter, the last she ever wrote, is a suffi-
" cient proof of the tenderness and composure of her mind.
" —Reader! it was written by a woman sinking from ease,
" youth, beauty, admiration, and love, to the grave."

" You do not, indeed, my dear friend, know how
" difficult it may be to write *one line*. I believe you have
" no idea of the manner of my life, which is, however, much
" more comfortable the last ten days, than some time pre-
" ceding.
" I have every blessing, and I am happy. The conver-
" sation of my beloved husband, when my breath will let
" me have it, is my greatest delight, he procures me every
" comfort, and as he always said he thought he should,
" contrives for me every thing that can ease and assist my
" weakness.

" Like a kind angel whispers peace,
" And smooths the bed of death.

" This part is not yet come, but I doubt not his steadi-
" ness and strong affection in those hours; as to the time

" of their arrival, I remain in the most perfect tranquillity
" and uncertainty. The spring is reckoned dangerous. If
" that time should decide the affair, Farewell, and let me
" intreat you, my dear friend, to believe, that I feel the
" highest respect for your virtues, and gratitude for the con-
" tinued warmth of your friendship.

" Adieu,
" HONORA EDGEWORTH."

This letter was written a few days before her death. It
was addressed to Mrs. Mary Powys.

CHAPTER XVI.

NOTHING is more erroneous than the common belief, that a man, who has lived in the greatest happiness with one wife, will be the most averse to take another. On the contrary, the loss of happiness, which he feels when he loses her, necessarily urges him to endeavour to be again placed in a situation, which had constituted his former felicity.

I felt, that Honora had judged wisely, and from a thorough knowledge of my character, when she had advised me to marry again, as soon as I could meet with a woman, who would make a good mother to my children, and an agreeable companion to me. She had formed an idea, that her sister Elizabeth was better suited to me than any other woman; and thought that I was equally well suited to her. Of all

Honora's sisters I had seen the least of Elizabeth. After she had declined my friend Mr. Day's addresses, I understood, that a gentleman, of a figure and manners uncommonly agreeable, was attached to her; and that she was not indifferent towards him. This gentleman had gone abroad, and it had been supposed, that, at some distant period, if he returned in circumstances sufficiently affluent, their marriage would be concluded. Notwithstanding this, Honora had spoken to her sister upon the subject then nearest to her heart. Miss Elizabeth Sneyd expressed the strongest surprise at the suggestion. Not only because I was her sister's husband, and because she had another attachment, but independently of these circumstances, as she distinctly said, I was the last man of her acquaintance, that *she* should have thought of for a husband; and certainly, notwithstanding her beauty, abilities, and polished manners, I believed that she was as little suited to me.

After some time I found the Sneyd family in great anxiety about her health, which was apparently declining, though she had no distinct symptoms of any disease. Dr. Darwin advised sea-bathing, and recommended it to her to go to Scarborough. Her brothers, Mr. E. and Mr. W. Sneyd, and her two sisters, Mary and Charlotte, were to go with her. They invited me and my children to accompany them. Miss Charlotte Sneyd's attention to my children, arising from her devoted affection to their mother, was the greatest possible inducement to my accepting this kind offer.

The journey was in some measure agreeable. None of my female companions had ever seen the sea; and I think, that I was more struck with the surprise, and admiration, which they expressed at the sight of the ocean, than at any circumstance which had occurred since the loss of my wife. My two brothers-in-law were extremely kind to me: the social, gentle temper of the one, and the excellent understanding of the other, combined in efforts to soothe and console me; their sisters then and ever shewed me the utmost friendship and attention; so that by degrees I was restored to myself, and to a sense of my situation.

Honora's advice pressed upon my mind; my thoughts fixed on Miss Elizabeth Sneyd. I learned, that the engagement, which had formerly subsisted, had been broken off; and gradually we both altered the opinion, that we were unsuited to each other; both influenced by the judgment and wishes of Honora.

It seldom happens, that people see with the eyes, or judge from the opinion of their very best friends, in choosing a companion for life. We were examples to the contrary. Unforeseen circumstances, however, interposed difficulties to our union; and certain officious friends produced a great deal of unnecessary vexation. The subject of this marriage became public, and was made an object of party disputes.

Many persons interfered, and in the Birmingham and other newspapers various replies and rejoinders appeared, which have sunk into oblivion.

Miss Elizabeth Sneyd's best friend, (out of her own family,) was Lady Holte, a woman of much knowledge of the world, and of great firmness of character. With this lady she had passed much of her youth; and now she was invited to her house in Cheshire, where at this period of our lives, when we most wanted it, we received from her, and from her daughter and son-in-law, Mr. and Mrs. Bracebridge, generous hospitality, and steady kindness.

After we had been asked three times in the parish church, we met to be married; but on the very morning appointed for our marriage the clergyman received a letter, which alarmed him so much, as to make me think it cruel to press him to perform the ceremony. Lady Holte took Miss Elizabeth Sneyd to Bath: I went to London with my children, took lodgings in Gray's-Inn-Lane, and had our banns published three times in St. Andrew's church, Holborn. Miss Elizabeth Sneyd came from Bath, and on Christmas day, 1780, was married to me in St. Andrew's church, in the presence of my first wife's brother, Mr. Elers, his lady, and Mr. Day.

We went immediately to my house at Northchurch, where we resided during part of the winter, till we had disposed of the lease of that house, and of our furniture. We then went to London, where our old friends shewed us much kindness.

Sir Joseph Banks was about this time (1780-1) president of the Royal Society. I was invited by him to become a member of that body; and I shortly afterwards contributed to its Transactions a paper on the resistance of the air to bodies of different shapes; a circumstance which had been suggested to me by some experiments recorded in Robins upon gunnery. I was still a member of the literary club I formerly mentioned. John Hunter* was our president; Sir George Shuckburgh, Sir William Hamilton, Captain Cook, and several other men of distinguished characters, were among us. Omai, a native of Otaheite, was then in England; and he might well have passed for an European. It was an extraordinary circumstance observable in his manners, that he had, after a few months' residence in England, more the air

---

* I fear there is an anachronism in the former pages about John Hunter and Sir George Shuckburgh.—Perhaps the circumstances I formerly mentioned did not happen till this time.

and appearance of a gentleman, than it would have been possible for a native of England, of ordinary station, to have obtained, after he had grown up to be a man*.

We passed the ensuing summer at Davenport Hall, in Cheshire, near Brereton Hall, the seat of Sir Charles Holte. We rented it from Mr. Davenport, to whom it belonged. He had entertained Rousseau there, when he was brought over from France by Hume. In one of the rooms there was an excellent picture of

---

Note by the Editor.

* It is remarkable, that Dr. Johnson made the same observation on the manner and appearance of Omai.

" He had been in company with Omai, a native of one of the South Sea islands, after he had been some time in this country. He was struck with the elegance of his behaviour, and accounted for it thus: ' Sir, he had passed his time, while in England, only with the best company; so that all he had acquired of our manners was genteel. As a proof of this, Sir, Lord Mulgrave and he dined one day at Streatham; they sat with their backs to the light, fronting me, so that I could not see distinctly; and there was so little of the savage in Omai, that I was afraid to speak to either, lest I should mistake one for the other."

Boswell's Life of Johnson.

the eccentric philosopher of Geneva. I believe that the print, which is prefixed to the English translation of his works, was taken from this picture. The people in the neighbourhood of Davenport, who had seen or spoken to him, thought him mad; perhaps they were not much mistaken.

Davenport Hall is a retired place, and nothing could be more uniform than our life, while we resided there. We walked, or read, or taught our children, during the morning; and we passed many of our evenings at Sir Charles Holte's magnificent house at Brereton. This house had been built in the time of Queen Elizabeth, when the style of building was in general convenient, and the rooms in good proportion.

In the dining room, which was spacious and lofty, the frieze was ornamented with the armorial bearings of every sovereign in the world, blazoned with perfect truth, and beautifully executed, with the coat of arms of the families into which they had intermarried. The escutcheons were of a moderate size. Altogether this had not only a general good effect, but an appearance of utility, which was satisfactory.

From the remote situation in which we lived, within reach but of few gentry, we had leisure to profit by the excellent library at Brereton Hall. For my own amusement also I procured a few tools, and executed some pieces of machinery, that were indeed more curious than important. I made a clock for the steeple at Brereton; and a chronometer of a singular construction, which I intended to present to the King; not from any peculiar merit, which it possessed as to accuracy, though it was sufficiently accurate for all common purposes; but I thought of presenting it, to add to his Majesty's collection of uncommon clocks and watches, which I had seen at St. James's. It went eight days, shewed the hours, minutes, and seconds on a common dial plate, by common hands, from one common centre; and yet it had no wheels, nor were any of its parts or movements connected by what is technically called *tooth and pinion.*

The power, by which the pendulum was kept in motion, was communicated from the weight, without any friction, except that of small pivots, none of which moved through a space of more than the eight hundredth part of an inch in one second. I mention this as a kind of mechanical paradox; not as a pursuit worthy the attention of any man, who can employ his time in something better than difficult trifles.

My mechanical occupations attracted the attention of some gentlemen in the neighbourhood, who told me, that there was a clergyman, who had a small living at no great distance, who was not only as fond of mechanics as myself, but who was also a man of much general knowledge. I requested, that my name should be mentioned to him, with an offer on my part to pay him a visit, if it would be agreeable. I had been informed, that since some family misfortune he never went from home. He returned an answer expressing a wish to see me, and I set out to find his residence, according to the directions I had received from the curate of the parish. After riding three or four miles through a well wooded country, I came suddenly upon the parsonage, at which this gentleman resided. The house was one of those irregular but convenient dwellings, which to an air of antiquity unite an appearance of neatness and comfort, of which sprucer and more modern buildings are often destitute. A white wicker gate opened into a small green court, the grass neatly mown, and nicely kept, with grapes hanging on the walls, and roses growing in profusion in the hedges.

All this was pleasing, yet not uncommon. But it was an uncommon sight, to behold an enormous white globe, erected upon an upright axis, in the middle of this enclosure, and round it several globes of less dimensions, the smallest of which was not larger than a pea. I readily comprehended, that these globes were intended to represent the relative sizes of the planets, their satellites, and their distances from each other; but I could not conceive how any man could have thought it worth his while, to execute such a laborious representation, unless he were a teacher of astronomy. I examined this apparatus leisurely before I went further, and was satisfied of its accuracy, and of the workmanlike manner in which it had been performed. At length I knocked at the house-door; and, after some time, it was opened

by a person in black, who, notwithstanding
great negligence in his dress, had the appearance of a gentleman. He had the remains of a fine person and countenance.
I told him my name, and he shewed me
into a parlour, which had been well furnished, but it was then a little decayed,
and littered with books and papers. He
made no apology for this disorder, but entered freely into conversation on various
subjects, and upon various books : he seemed pleased, and, after some time, I ventured to inquire the purpose of the machinery, which I had seen before his door.

His countenance immediately changed,
and, speaking with great agitation, he told
me, that the globes which I had noticed
were the work of his son, a boy of fourteen,
whose talents, acquirements, and industry
were far beyond his years; and who was
as good as he was ingenious. The father
added, that as he had lost this boy's mother, the youth had been for some years his
only solace: that two years ago he had
been deprived of this promising and affectionate son, by a lingering disease. He
said, that he had been his boy's only mas-

ter; and that, beside the learned languages, he had given him a taste for English literature, and some knowledge of mathematics, and of natural philosophy. These instructions had been eagerly seized. Every hour, which was not dedicated to books, had been usefully employed; so that, at his very early age, the boy had been able to execute many philosophical and optical instruments, and to give some adequate notion of the sizes of the planets. For the amusement and instruction of his young companions, he had, with his own hands, and with indefatigable industry, constructed the globes, which I had seen. The large ones were made of lath and plaister; the smaller of plaister of Paris.

I endeavoured, for some time in vain, to turn the father's thoughts aside from this subject, as I saw that it threw him into violent agitation. He still recurred to the object nearest his heart; and continually adduced fresh instances of his son's abilities, and of his amiable disposition.

It struck me, that his grief had injured his understanding, and I felt myself in a very painful situation; but at length the door opened, and a girl of fourteen or fifteen entering with a tea-tray in her hand suddenly stopped the course of his thoughts. He looked at her with such complacency, or rather with such admiration, as might be expected from a person, who had, for the first time, beheld a most beautiful object; and such indeed she was! Her dress was white, and was fashioned more like the dresses now in use, than according to the custom of the times when I saw her: she had none of those stiff and cumbersome bodices, which girls of her rank then usually wore. The natural shape appeared; her shoulders were partly uncovered, her bosom but modestly revealed; what could be seen was not of that dead white, which poets compare with driven snow, but of that healthful whiteness, which admits of no comparison with any inanimate object in the creation; her limbs and feet were beautifully shaped, her hair hung in a profusion of ringlets down her neck; the hair was of that peculiar auburn, that has no appearance of yellow but what is golden, and no tint approaching to red. Her features were not exactly regular, her blue eyes and aquiline nose were larger in proportion than her mouth; her eye-brows darker than her hair, and the contour of her face so exactly between flatness and too great prominence, as would have baffled the pencil to delineate. In a word, such perfect beauty I had never seen before. The expression of her countenance, the grace of her motion, unspoiled by masters or by affectation, and the sound of her harmonious voice, altogether amazed me.

" And she is better than she is beautiful, and as wise as she is good!" exclaimed my host, as she withdrew.—" She is a scholar, and an artist, and is neither proud nor vain of any of her mental or personal accomplishments."

We sat down to tea, however, without the young person. She appeared once or twice as a vision. To my imagination she resembled nothing that was real; and, in ancient fable, I could compare her to nothing ideal but the goddess Hebe. Her youth, the innocence and perfect modesty of her deportment, instantly destroyed any supposition, that she was his mistress. But, if she was his daughter, why did not she preside at the tea-table?

After tea he took me up stairs, into a

gallery filled with excellent books, not in handsome bindings, or accurately arranged, but all well chosen; and it was obvious that they had been often used. Here, without any ostentatious display, he proved to me, that his mind was stored with modern literature and classical learning. He told me, that, during the life of his son, this library was a world to him; that his situation was too humble to give him access to the great, or the highly educated, and that he had no temptation to mingle with the vulgar and the ignorant.

After a conversation that detained me to a late hour, I took my leave with a mixture of pleasure and regret. To meet with such abilities and information, in such perfect seclusion, excited pleasure and surprise. To see such talents and acquirements totally lost to society, and nearly useless to their possessor, necessarily excited compassion and disappointment.

I scarcely ventured to make any farther inquiries about this gentleman: I feared that something was not right. But from what little I did hear, the fact seemed to be, that the mother of his favourite son had not always been his wife ; and that his pecuniary difficulties arose from his being entangled in the meshes of the law by a nefarious attorney, who confined him to his house by writs, which were not legally obtained. This attorney had impoverished him by the most vexatious litigation, and had vowed that his life should finish in a jail.

It was, perhaps, fortunate for me, that I shortly afterwards left that neighbourhood.

I could not have refrained from making some attempt to deliver him, which might have involved me in much difficulty, and which in the end would probably have been ineffectual.

Except this incident, scarcely any thing worth mentioning happened to me at Davenport.

A short time before I left Davenport, I became acquainted with * * * * * *
* * * * * * * * * * * * * * * * * *
* * * * * * * * * * * * * * * * *

END OF VOL. I.

J. M'Creery, Printer,
Black-Horse Court, London.

Printed in the United States
By Bookmasters